TABLE OF CONTENTS

OPTICAL FIBER SENSORS

1988 TECHNICAL DIGEST SERIES, VOLUME 2, PART 2

Conference Edition

Summaries of papers presented at the Optical Fiber Sensors Topical Meeting

January 27–29, 1988
New Orleans, Louisiana

Cosponsored by

Optical Society of America
Lasers and Electro-Optics Society of the
Institute of Electrical and Electronics Engineers

Optical Society of America
1816 Jefferson Place, N.W.
Washington, D.C. 20036
(202) 223-8130

Articles in this publication may be cited in other publications. In order to facilitate access to the original publication source, the following form for the citation is suggested:

Name of Author(s), Title of Paper, Optical Fiber Sensors, 1988 Technical Digest Series, Vol. 2, (Optical Society of America, Washington, D.C. 1988) pp. xx–xx.

ISBN Number
Conference Edition 1-55752-021-6 (softcover; Part 1)
Conference Edition 1-55752-036-4 (softcover; Part 2)
Postconference Edition 1-55752-022-4 (hardcover; Parts 1 and 2)
(Note: Postconference Edition includes postdeadline papers.)

Library of Congress Catalog Card Number
Conference Edition 87-062764 (Parts 1 and 2)
Postconference Edition 87-062765 (Parts 1 and 2)

IEEE Catalog Number 88CH2524-7

WEDNESDAY, JANUARY 27, 1988

MEETING ROOM 2-4-6

8:30–10:00am
WAA, Fiber Sensors Plenary Session
Shaoul Ezekiel, *Massachusettes Institute of Technology, Presider*

8:30am **WAA1 Fiber-Optic Sensors for Sale?** A. L. Harmer, *Battelle Geneva, Switzerland.* The fiber-optic sensor market, unlike the fiber-optic communications market, is highly heterogeneous and diversified in both sensor types and uses. This paper reviews the many functions of sensors, their basic technology, performance, construction, and end user markets. (p. 2) **(Plenary paper)**

9:00am **WAA2 Biomedical Fiber-Optic Sensors,** Abraham Katzir, *Tel Aviv U., Israel.* Fiber-optic sensors can be used for monitoring physical (P, T, flow) or chemical (pH, pO_2, glucose content) quantities. These sensors may be inserted into the body via catheters or endoscopes, and used as diagnostic tools. They are potentially sensitive, reliable, and cost effective. (p. 4) **(Plenary paper)**

9:30am **WAA3 Ocean Applications for Fiber-Optic Sensors,** Barry E. Paton, *Dalhousie U., Canada.* The oceans provide a unique arena where many of the attributes of optical sensing can be and are being used today. However, the oceans present many challenges in using fiber-optic sensors in this especially hostile environment. (p. 7) **(Plenary paper)**

WEDNESDAY, JANUARY 27, 1988—*Cont.*

MEETING ROOM 2-4-6

10:30am–12:00m
WBB, Special Purpose Fibers
Juichi Noda, *NTT Electrical Communications Laboratory, Japan, Presider*

10:30am **WBB1 Coatings for Optical Fibers,** James H. Cole, *Dylor Corporation.* The development of coatings for optical fibers is reviewed. Standard fiber coatings as applicable for the telecommunications industry are compared with specialized coatings required for use in fiber sensors. (p. 10)**(Invited paper)**

11:00am **WBB2 Mode Characteristics of Highly Elliptical Core Two-Mode Fibers Under Perturbations,** *S. Y. Huang, Shanghai Jiao Tong, U. China;* J. N. Blake, B. Y. Kim, *Stanford U.* The modal and polarization behavior of a highly elliptical core two-mode fiber under various external perturbations such as axial strain, temperature change, twist, squeezing and bending, is investigated. (p. 14)

11:15am **WBB3 Polarization-Maintaining Fibers for Fiber Optical Gyroscopes,** E. Sasaoka, H. Suganuma, S. Tanaka, *Sumitomo Electric Industries, Ltd., Japan.* Polarization-maintaining fiber with extremely small coating diameter, 133 μm, has been developed for a compact fiber optical gyroscope. A 1000-m long fiber coil of 65-mm diameter showed crosstalk as low as -14.4 dB. (p. 18)

11:30am **WBB4 Environmental Testing of Small Diameter Polarization-Maintaining Fiber Coils for Optical Sensor Systems,** Dan Courtney, Schuyler Montgomery, Gary Boivin, Tim Bailey, *Hamilton Standard.* Results of environmental testing of small diameter fiber coils are given. Measurements of optical loss and *h*-parameter were made over a temperature range of -60 to $+60$ °C. Performance of a fiber gyro using a small coil with applied temperature gradients is given. Winding apparatus for fiber coils is also discussed. (p. 25)

11:45am **WBB5 Lapped Polarization-Maintaining Fiber Resonator,** R. Dahlgren, *Honeywell Inc.* Results of an effort to develop lapped polarization-maintaining fiber-optic couplers and resonators are presented. High performance optical-contact-bonded spliceless resonators with high polarization extinction ratio, finesse, and modulation depth are described. (p. 29)

MEETING ROOM 2-4-6

1:30–3:00pm
WCC, Interferometric Techniques
Luc B. Jeunhomme, *Photonetics, France, Presider*

1:30pm **WCC1 Heterodyne Technology for Optical Sensors,** Toshihiko Yoshino, *Tokyo U., Japan.* With heterodyne technology such quantities as temperature, pressure, strain, displacement, rotation (gyro), magnetic and electric fields, current, film thickness and optical constants can be measured with high precision, high stability, large dynamic range, and also at distant points. Sensing schemes and typical performances are presented. (p. 40)**(Invited paper)**

2:00pm **WCC2 Elimination of Polarization-Induced Signal Fading in Interferometric Fiber Sensors Using Input Polarization Control,** A. D. Kersey, *Sachs/Freeman Associates;* A. Dandridge, A. B. Tveten, *U.S. Naval Research Laboratory.* We present an analysis of polarization-induced fading in interferometric fiber sensors which predicts that the effect can always be overcome through control of the state of polarization of the light input to the interferometer. Experimental results verifying this theory are reported. (p. 44)

2:15pm **WCC3 Interferometric Sensors Using Internal Fiber Mirrors,** C. E. Lee, H. F. Taylor, *Texas A&M U.* Reflectively monitored Fabry-Perot and Michelson interferometers which make use of dielectric mirrors in continuous lengths of single-mode fiber are characterized for use as temperature and wavelength sensors. (p. 48)

2:30pm **WCC4 Elimination of Intensity-Induced Sideband Noise in Interferometric Fiber Sensors,** F. Bucholtz, M. J. Marrone, A. Dandridge, *U.S. Naval Research Laboratory;* K. P. Koo, M. A. Davis, *Sachs/Freeman Associates.* Upconversion of low-frequency optical intensity noise can degrade the signal-to-noise ratio (SNR) in sensors employing phase carriers. We demonstrate a 20–25-dB improvement in the SNR in the presence of a 300-mrad rms phase carrier by stabilizing the laser intensity. (p. 52)

2:45pm **WCC5 Short Coherence Length Interferometric Fiber-Optic Sensor System,** A. S. Gerges, F. Farahi, T. P. Newson, J. D. C. Jones, D. A. Jackson, *U. Kent, U.K.* A fiber-optic sensor system exploiting the principles of white light interferometry is demonstrated. A digital feedback servo ensures that the Mach-Zehnder interferometric system is maintained at near zero optical path difference. (p. 56)

WEDNESDAY, JANUARY 27, 1988—*Cont.*

MEETING ROOM 2-4-6

3:30–5:30pm
WDD, Multiplexed and Distributed Sensors
Kazuo Kyuma, *Mitsubishi Electric Corporation, Japan, Presider*

3:30pm **WDD1 Multiplexed and Distributed Optical Fiber Sensors,** A. Dandridge, *U.S. Naval Research Laboratory;* A. D. Kersey, *Sachs/Freeman Associates.* The principles involved in multiplexed and distributed optical fiber sensors are discussed, and the latest theoretical and experimental developments in these areas reviewed. (p. 60)
(Invited paper)

4:00pm **WDD2 Coherent Backscatter-Induced Excess Noise in Reflective Interferometric Fiber Sensors,** A. M. Yurek, A. Dandridge, *U.S. Naval Research Laboratory;* A. D. Kersey, *Sachs/Freeman Associates.* We discuss the effects of coherent backscatter-induced excess intensity noise in remotely interrogated reflective-type interferometric sensor configurations such as the Michelson and low-reflectance Fabry-Perot. Experimental results demonstrating the noise mechanism are presented. (p. 72)

4:15pm **WDD3 Frequency Division Multiplication of Optical Fiber Sensors Using an Optical Delay Line with a Frequency Shifter,** Der-Tsair Jong, Kazuo Hotate, *U. Tokyo, Japan.* We propose a new method for multiplexing a fiber sensor array. In the system, the input signals can be multiplexed/demultiplexed by using a series of heterodyne outputs with different intermediate frequencies. (p. 76)

4:30pm **WDD4 Tapped Serial Interferometric Fiber Sensor Array with Time Division Multiplexing,** A. D. Kersey, *Sachs/Freeman Associates;* A. Dandridge, *U.S. Naval Research Laboratory.* We report a new architecture for multiplexed interferometric fiber-optic sensors based on time-division addressing. Intrinsic crosstalk of the system is analyzed and compared to experimental results obtained using a three-element network. (p. 80)

4:45pm **WDD5 At-Sea Deployment of a Multiplexed Fiber-Optic Hydrophone Array,** M. L. Henning, C. Lamb, *Plessey Naval Systems, U.K.* A six-hydrophone serially multiplexed array driven by a single laser has been deployed at sea. Its performance has been monitored over an extended period, and compares with equivalent piezo-electric systems. (p. 84)

5:00pm **WDD6 Distributed Liquid Sensor Using Eccentrically Cladded Fiber,** Hiroshi Yoshikawa, Minoru Watanabe, Yutaka Ohno, *Nihon U., Japan.* A 420-m long distributed liquid sensor is demonstrated. Consisting of a single-mode fiber with an eccentric core, the sensor can detect presence and positions of liquids by using an optical time-domain reflectometer. (p. 92)

5:15pm **WDD7 5-mm Resolution Optical Frequency Domain Reflectometry Using a Coded Phase Reversal Modulator,** Moshe Nazarathy, David W. Dolfi, Steven A. Newton, *Hewlett-Packard Laboratories.* A 5-mm resolution OFDR technique was demonstrated over a 20-GHz frequency span using a 13-bit Barker coded phase reversal optical modulator. (p. 96)

THURSDAY, JANUARY 28, 1988

MEETING ROOM 2-4-6

8:00–10:15am
ThAA, Magnetic, Electrical, and Displacement Sensors
Yoshito Ueno, *Nippon Electric Company, Japan, Presider*

8:00am **ThAA1 Distributive Effects in a Fiber-Optic Magnetostrictive Transducer Using Metallic Glass,** K. P. Koo, F. Bucholtz, A. Dandridge, *U.S. Naval Research Laboratory.* The response of a fiber-optic magnetostrictive transducer using metallic glass has been correlated to the behavior of a distributive transducer. (p. 106)

8:15am **ThAA2 Optoelectronic Processing for Faraday Effect Magnetometry: Applications for Remote Measurement,** R. P. Tatam, M. Berwick, J. D. C. Jones, D. A. Jackson, *U. Kent, U.K.* Faraday effect magnetometry using a rotating linear polarization state to enable pseudoheterodyne processing is described. Increased sensitivity through the use of a higher Verdet constant sensing element and the feasibility of remote operation are discussed. (p. 110)

8:30am **ThAA3 Design of a Totally Dielectric Fiber-Optic rf Electric Field Sensor,** Mark L. Wilson, Dan J. Bartnik, Mark P. Bendett, *Unisys Corp.* This paper details research to develop a fiber-optic sensor to detect radio frequency electric fields. The sensor was constructed using a custom polyvinylidene fluoride-coated D-shaped optical fiber. Fabrication techniques and test results are discussed. (p. 114)

8:45am **ThAA4 Fiber-Optic Magnetic Field/Current Sensors Utilizing Lorentzian Force,** Haruo Okamura, *NTT Transmission Systems Laboratories, Japan.* Fiber-optic magnetic field/current sensors utilizing Lorentzian force are reported with sensitivities of 26 rad/Gs • 100 mm and 555 rad/A • 100 mm. These dc sensors can be used for simultaneous multipoint measurements by frequency superimposition. (p. 118)

9:00am **ThAA5 Compact Optical Fiber Current Monitor with Passive Temperature Stabilization,** R. I. Laming, D. N. Payne, L. Li, *U. Southhampton, U.K.* Quasicircularly birefringent fiber and a broad-spectrum light source are combined to obtain an accurate, compact and robust current monitor. Measurement repeatability of $\pm 0.5\%$, a temperature drift of 0.05%/°C, and a sensitivity of 1 mA rms/$HZ^{0.5}$ characterize the sensor. (p. 123)

9:15am **ThAA6 Distance Sensing Using Constant-Amplitude Wavelength-Swept Semiconductor Lasers,** E. M. Strzelecki, D. A. Cohen, L. A. Coldren, *UC–Santa Barbara.* Using electronically tunable constant-power two-section semiconductor lasers and a coherent receiver in an all-fiber-optic interferometer, we have investigated the range limits, accuracy, and resolution of a noncontact optical distance sensor. (p. 129)

9:30am **ThAA7 Line Loss Independent Fiber-Optic Displacement Sensor with Electrical Subcarrier Phase Encoding,** S. Ramakrishnan, L. Unger, R. Kist, *Fraunhofer-Institut fur Physikalische Messtechnik, F.R. Germany.* A line loss insensitive macrodisplacement sensor with electrical subcarrier phase encoding using optical fiber bundle dividers has been developed. The principle of the method and the characteristics of the sensor as measured for displacements up to a few hundred millimeters are presented. (p. 133)

THURSDAY, JANUARY 28, 1988—*Cont.*

9:45am **ThAA8 Integrated-Optic Microdisplacement Sensor Using a Y Junction and a Polarization-Maintaining Fiber,** Hayami Hosokawa, Junichi Takagi, Tsukasa Yamashita, *Omron Tateisi Electronics Co., Japan.* An integrated-optic microdisplacement sensor has been developed. Simultaneous detection of the direction and the quantity of displacement were achieved by use of the phase difference between TE and TM modes. (p. 137)

10:00am **ThAA9 Fiber-Optic Position Sensor Array,** M. Brenci, G. Conforti, A. G. Mignani, Annamaria Scheggi, *IROE-CNR, Italy.* An optical head made up of four GRIN lens-ended optical fibers giving spatial information over four sensing zones has been designed and tested. Two geometrical arrangements of the position of the lenses and their axis inclinations are proposed; experimental results are discussed. (p. 141)

MEETING ROOM 2-4-6

10:30am–12:00m
ThBB, Fiber Components
Mark Johnson, *York Harburg Sensor GmbH, F. R. Germany, Presider*

10:30am **ThBB1 Few-Mode Fiber Devices,** B. Y. Kim, *Stanford U.* A new class of fiber-optic devices using optical fibers that support more than one spatial mode is described. The use of highly elliptical core two-mode optical fiber in sensors and passive and active guided wave components are discussed. (p. 146) **(Invited paper)**

11:00am **ThBB2 Resonant-Loop Optical Fiber Phase Modulator,** M. N. Zervas, I. P. Giles, *U. College London, U.K.* A novel phase modulator, consisting of a loop of optical fiber attached to a piezoelectric plate, is described. Resonance frequency tuning and tailored frequency response is achievable with low temperature drift and minimal polarization modulation. (p. 150)

11:15am **ThBB3 Phase Modulation in an All-Fiber Ring Resonator Sensor,** Z. K. Ioannidis, P. M. Radmore, I. P. Giles, *U. College London, U.K.* The effect of high frequency phase modulation and nonlinear phase ramps on an all-fiber resonator is theoretically and experimentally described. This has implications in phase bias control systems and high frequency sensing. (p. 154)

11:30am **ThBB4 All-Fiber Acousto-Optic Phase Modulators Using Zinc Oxide Films on Glass Fiber,** A. A. Godil, D. B. Patterson, B. L. Heffner, G. S. Kino, B. T. Khuri-Yakub, *Stanford U.* An advanced design of an all-fiber acoustooptic phase modulator is demonstrated using zinc oxide films on standard single-mode fiber, in the 450-MHz frequency range, with gallium to provide heat sinking and acoustic backing. (p. 159)

11:45am **ThBB5 In-Fiber Bragg-Grating Sensors,** G. Meltz, W. W. Morey, W. H. Glenn, J. D. Farina, *United Technologies Research Center.* Bragg-grating sensors have been formed in the core of a germanosilicate fiber by lateral exposure to a 257-nm two-beam interference pattern. The technique offers a new means for quasidistributed temperature and strain measurements. (p. 163)

THURSDAY, JANUARY 28, 1988—*Cont.*

EXHIBIT HALL C

1:00–3:00pm
ThCC, OFS Poster Session

11:00am–1:00pm
Poster Preview

1:00pm–3:00pm
Authors Present

ThCC1 Optical Heterodyne Gyroscope Using Two Reversed Fiber Coils, H. Koseki, Y. Imai, Y. Ohtsuka, *Hokkaido U., Japan.* A new heterodyne fiber-optic gyroscope is devised. Two optical waves circulating in the same direction along the reversed coils yield a beat-photocurrent signal at both ends. Their phase difference gives the rotation rate. (p. 168)

ThCC2 Piezoelectric Crystal Based Fiber-Optic Pressure Sensor, Michael H. Ikeda, Mei H. Sun, *Luxtron Corp;* Stephen R. Phillips, *Alamo Instruments.* A fiber-optic pressure sensor based on a piezoelectric quartz crystal is described, utilizing the pressure-dependent nature of the oscillation frequency of the crystal. A detection sensitivity of 4-mm Hg pressure has been demonstrated. (p. 172)

ThCC3 Differential Thermal Analysis Using an Optical Fiber Interferometer, T. Kimura, N. Kobayashi, S. Takahashi, O. Akimoto, Ring Ping, K. Noda, *Tokyo U. Agriculture & Technology, Japan.* A temperature difference measuring system using a fiber-optic interferometer is used in differential thermal analysis of various physical or chemical small heat phenomena. The minimum detectable heat is 75.5 μcal for 3 cc of material. (p. 179)

ThCC4 Relativistic Electron Beam Measurement by Cherenkov Radiation Generated in an Optical Fiber, Tomonori Oie, *Diesel Kiki Co., Japan;* Ken-Ichi Ueda, Hiroshi Takuma, *UEC Institute for Laser Science, Japan.* Generation of Cherenkov radiation in a silica optical fiber by a relativistic electron beam is investigated as a new method of observing temporal and spatial distribution of intense (250–600 keV, 100–400 A/cm^2) electron beams. (p. 184)

ThCC5 Double-Polarization Interferometer for Digital Displacement and Force Sensing by Fiber Tension-Bending, Norbert Furstenau, *German Aerospace Research Establishment, F.R. Germany.* An interferometric displacement sensor with digital readout by fringe counting is investigated. Displacement is measured via tension bending induced fiber strain. Ambiguity in fringe counting is eliminated employing the double-polarization method. (p. 191)

ThCC6 Highly Selective Single-Mode Fiber-Optic Couplers, A. G. Bulushev, Ju. V. Gurov, E. M. Dianov, O. G. Okhotnikov, V. M. Paramonov, A. M. Prokhorov, *Academy of Sciences of the U.S.S.R.* Fused fiber-optic couplers with a spectral selectivity up to 7×10^{-2} nm, excess losses < 1 dB, and an extinction ratio of > 20 dB have been demonstrated for the first time. All-fiber ring resonators with selective couplers have been considered. (p. 195)

ThCC7 Interferometric Optical Fiber Strain Sensor Under Biaxial Loading, Jim Sirkis, *U. Florida.* The analysis and experimental verification of a novel fiber-optic strain sensor is presented. This interferometric strain sensor uses an S-shaped sensing fiber and a circular shaped reference fiber. This geometry leads to a linear relation between the strain in the *x*-direction and the change in phase. (p. 203)

ThCC8 Novel Frequency-Out Optical Fiber Sensing Technique, M. V. Andres, *U. Valencia, Spain;* W. K. H. Foulds, M. J. Tudor, *U. Surrey, U.K.* An optical fiber resonant ring is used to detect the flexural vibration modes of a particular section of the ring, which effectively becomes a resonant sensor capable of measuring a number of physical quantities. (p. 208)

ThCC9 Optical Fiber Strainmeter and Test of Its Stability, Mark Zumberge, Frank Wyatt, Don X. Yu, *UC-San Diego.* We have built two optical fiber strainmeters at our research facilities in southern California. Using single-mode optical fiber, we measure strain of the fiber interferometrically. We have also carried out stability experiments on the fibers over long periods of time (months to years). (p. 212)

ThCC10 Low-Loss Highly Overcoupled Fused Couplers: Fabrication and Sensitivity to External Pressure, F. Bilodeau, K. O. Hill, S. Faucher, D. C. Johnson, *Department of Communications, Canada.* A novel flame brush technique is used to fabricate optical directional couplers that maintain low loss over many coupling cycles. Such couplers are shown to be very sensitive to externally applied hydrostatic pressure. (p. 221)

ThCC11 Heterodyne Polarimetric Detection for Remote Sensing, R. Calvani, R. Caponi, F. Cisternino, *CSELT, Italy.* An original heterodyne polarimeter for remote sensing is presented. Good performance in high frequency operation and resolution close to the shot noise limit are demonstrated. Heterodyne recovery of weak polarization signals and frequency multiplexing of sensors are feasible. (p. 226)

ThCC12 Optical Fiber Angular Displacement Sensor Employing the Surface Plasmon Resonance Effect, C. M. France, B. E. Jones, *Brunel U., U.K.* A passive angular-displacement sensor with optical fiber links using absorbance wavelength modulation has been demonstrated. A simple prism and thin film are employed which exhibit surface plasmon resonance. Test results cover an 18° angular displacement range with a resolution of ~0.5°, 8-dB insertion loss, and temperature sensitivity of ~0.01/°C. (p. 234)

ThCC13 Nulling Coherent Backscatter in Optical Fiber Gyroscopes, Brian Culshaw, James M. Mackintosh, *U. Strathclyde, U.K.* A brief theoretical analysis precedes new experimental results which demonstrate that control of the input coupler ratio and the detection phase enable the backscatter signal to be totally separated from the Sagnac phase. (p. 240)

THURSDAY, JANUARY 28, 1988—*Cont.*

ThCC14 Fiber-Optic Rotary Displacement Sensor with Wavelength Encoding, W. B. Spillman, Jr., *Simmonds Precision;* P. L. Fuhr, *U. Vermont.* A fiber-optic sensor is described in which an optical retardation plate is used to encode rotary displacement information as a notched minimum in a broadband optical signal. As the waveplate is rotated, the optical beam experiences a variable linear retardation. The signal wavelength at which the retardation is exactly one-half wave exhibits a minimum intensity transmission. The wavelength of the intensity minimum is then a function of the rotation of the retardation plate. Theoretical prediction of the sensor's performance is developed and compared to experimental results. (p. 244)

ThCC15 Optical Fiber Fabry-Perot Temperature Sensors, Shiao-Min Tseng, Chin-Lin Chen, *Purdue U.* A new and simple family of optical fiber Fabry-Perot temperature sensors capable of measuring temperature change and discerning the temperature rise from the temperature drop is proposed and demonstrated. (p. 252)

ThCC16 Single-Mode Coupler Sensors, Timothy E. Clark, Mary W. Hall, *McDonnell Douglas Astronautics Co.* We have constructed sensors from single-mode couplers by altering the final package to enhance desired responses. Devices made include a temperature insensitive bending sensor and a thermal switch. (p. 256)

ThCC17 Two-Photon Absorption of Excimer-Laser Radiation in Optical Fibers, Toru Mizunami, Keiji Takagi, *Kyushu Institute of Technology, Japan.* Two-photon absorption in UV-grade optical fibers is measured with a XeBr excimer laser (282 nm). The two-photon absorption coefficient is $5 \pm 1 \times 10^{-6}$ cm/MW. A theoretical value based on the Keldysh formula is discussed. (p. 260)

ThCC18 Analysis of a Novel Optical Fiber Interferometer with Common-Mode Compensation, A Bruce Buckman, *U. Texas at Austin.* A novel fiber interferometer, formed by feeding back the unused output of a Mach-Zehnder through a length of fiber to the unused input port, is shown to compensate for environmental and wavelength fluctuations. (p. 268)

ThCC19 Electronic Interferometric Sensor Simulator/Demodulator, A. B. Tveten, E. C. McGarry, A. Dandridge, *U.S. Naval Research Laboratory;* A. D. Kersey, *Sachs/Freeman Associates.* We describe a simple interferometric sensor simulator circuit which has a wide range, ultra-low noise, and a wide bandwidth. The device is useful in the testing/characterization of sensor demodulation electronics, and can also be configured as a phase-tracking sensor demodulator. (p. 277)

ThCC20 Modal Analysis and Design Criteria of Azimuthally Inhomogeneous Circularly Birefringent Fibers, Carlo G. Someda, Roberto Castelli, Fernanda Irrera, *U. Padua, Italy.* We present (i) exact analytical modes of a family of azimuthally inhomogeneous fibers; (ii) examples showing how the fiber parameters affect field distributions, dispersion relationships, cut-off wavelengths; (iii) criteria for aiming at high circular birefringence or at low input mismatch loss. (p. 281)

ThCC21 High Accuracy Faraday Rotation Measurements, Edward A. Ulmer, Jr., *Square D Co.* A simple, new, passive method to measure Faraday rotation with high accuracy in the presence of time-varying birefringence is derived from first principles and used in fiber-optic electric current sensor. (p. 288)

ThCC22 Magnetic-Field Sensor with Polarization-Maintaining Optical Fiber, F. Suzuki, Y. Kikuchi, T. Shiota, O. Fukuda, K. Inada, *Fujikura, Ltd., Japan.* A new magnetic-field sensor composed of polarization-maintaining fibers and a distributed feedback semiconductor laser diode have been fabricated. A minimum detectable magnetic field of 3×10^{-6} Oe/fiber•m is demonstrated. (p. 292)

ThCC23 Two Fiber-Optic Accelerometers, C. J. Zarobila, J. B. Freal, R. L. Lapman, C. M. Davis, *Optical Technologies, Inc.* Two fiber-optic accelerometers have been designed, fabricated, and tested. Their design is based on simple spring-mass motion. For each device, the minimum detectable accelerations depend on case displacement and design mechanical resonance. (p. 296)

THURSDAY, JANUARY 28, 1988—*Cont.*

MEETING ROOM 2-4-6

1:30–3:15pm
ThDD, Integrated-Optic Devices
Masamitsu Haruna, *Osaka University, Japan, Presider*

1:30pm **ThDD1 Passive Integrated-Optic Devices,** R. T. Kersten, *Schott Glass Works, F.R. Germany;* W. Ross, *Integrated Optics Technology, F.R. Germany.* Fiber- and integrated-optic sensors are gaining importance. For the new generation of integrated- and fiber-optic sensors the micro- and fiber-optic elements will be replaced by integrated optics. IO technologies and uses of optical sensors and IO-elements are discussed. (p. 304)
(Invited paper)

2:00pm **ThDD2 Active Integrated Optics for Sensors,** Masamitsu Haruna, Hiroshi Nishihara, *Osaka U., Japan.* Integrated-optic technology can provide compact and rugged fiber-optic sensors. We present our recent progress in $LiNbO_3$ optical ICs for the fiber laser Doppler velocimeter. In particular, a new optical IC for time-division 2-D velocity measurement is proposed. (p. 305)
(Invited paper)

2:30pm **ThDD3 Ti-$LiNbO_3$ Waveguide Serrodyne Modulator with Ultrahigh Sideband Suppression for a Fiber-Optic Gyroscope,** H. Hung, C. Laskoskie, T. El-Wailly, C. L. Chang, *Honeywell, Inc.* We report a Ti-$LiNbO_3$ waveguide serrodyne modulator with overall sideband suppression of 55 db. The phase modulator, driven by a linear ramp signal, provides such ultrahigh suppressions for optical frequency shifting of up to 200 kHz. (p. 309)

2:45pm **ThDD4 Fiber-Optic Interferometer with Digital Heterodyne Detection Using Lithium Niobate Devices,** D. Eberhard, *Fraunhofer Institute, F.R. Germany;* E. Voges, *U. Dortmund, F.R. Germany.* Michelson and Mach-Zehnder interferometers are built-up with polarization-maintaining fibers and Lithium niobate integrated-optic devices. Heterodyne detection by digital phase modulation provides accurate and programmable detection schemes. (p. 313)

3:00pm **ThDD5 High Quality Integrated-Optic Polarizers in $LiNbO_3$,** P. G. Suchoski, T. K. Findakly, F. J. Leonberger, *United Technologies Research Center.* High quality proton-exchanged integrated-optic polarizers are reported in Ti:$LiNbO_3$ waveguides. The polarizers, which operate at 0.82 μm, exhibit a polarization extinction of 55–60 dB and an excess insertion loss of 0.2 dB. (p. 317)

THURSDAY, JANUARY 28, 1988—*Cont.*

MEETING ROOM 2-4-6

3:30–5:00pm
ThEE, Sensor Components
Mokhtar S. Maklad, *EOtec Corporation, Presider*

3:30pm **ThEE1 Lead-Insensitive Optical Fiber Sensors,** John P. Dakin, *Plessey Research (Roke Manor) Ltd., U.K.* We describe the methods by which optical fiber sensors, intended for remote measurement, may be rendered insensitive to physical perturbation of the sensor leads and to any changes that could otherwise occur due to remating of connectors or lead replacements. (p. 323) **(Invited paper)**

4:00pm **ThEE2 Surface Plasmon Resonances in Thin Metal Films for Optical Fiber Devices,** G. Stewart, W. Johnstone, Brian Culshaw, T. Hart, *U. Strathclyde, U.K.* Simple expressions describing surface plasmon resonances in thin metal films are presented. Results are confirmed experimentally leading to the design of optical fiber polarizers. Further important applications for sensor and nonlinear devices are discussed. (p. 328)

4:15pm **ThEE3 Wavelength Conversion for Enhanced Fiber-Optic Ultraviolet Sensing,** R. A. Lieberman, L. G. Cohen, V. J. Fratello, E. Rabinovich, *AT&T Bell Laboratories.* We analyze the use of fluorescent phosphor-tipped detector fibers to increase the effective range of optical fibers used for remote ultraviolet sensing, and present two methods for using phosphors with fibers. (p. 332)

4:30pm **ThEE4 Optically Powered Sensors,** Paul Bjork, James Lenz, Kyuri Fujiwara, *Honeywell Systems & Research Center.* The design and experiments of a single fiber data link are described. Optical energy and control signals are transmitted to the sensor to electrically power sensor electronics and an LED data signal from the sensor is returned. (p. 336)

4:45pm **ThEE5 Garnet-Based Optical Isolator for 0.8-μm Wavelength,** G. L. Nelson, J. M. Sittig, C. H. Herbrandson, *Unisys Corp.* An optical isolator was constructed which has 2.7-dB insertion loss and 35 dB of isolation at 0.786-μm wavelength. Garnet films with 0.8, 1.3, 1.9, and 2.1 Bi/formula were grown and the 0.8 Bi/formula films included garnet antireflection layers. (p. 340)

FRIDAY, JANUARY 29, 1988

MEETING ROOM 2-4-6

8:00–10:00am
FAA, Chemical–Biological Sensors
Annamaria Scheggi, *Instituto de Ricerca Sulle Onde Electromagnetiche, Italy, Presider*

8:00am **FAA1 Distributed Fluorescence Oxygen Sensor,** R. A. Lieberman, L. L. Blyler, L. G. Cohen, *AT&T Bell Laboratories.* A distributed oxygen sensor, based on evanescent coupling of fluorescence to guided modes, has been fabricated by cladding silica fiber during the draw with dye-doped silicone. Oxygen sensitivity, response time, and lifetime have been investigated. (p. 346)

8:15am **FAA2 Design and Evaluation of a Reversible Fiber-Optic Sensor for Determination of Oxygen,** Marion R. Surgi, *Allied-Signal Engineered Materials Research Center.* The oxygen sensor developed has a useful second order Stern-Volmer response to gaseous oxygen concentrations between 0.5 and 100%. The evidence indicates that the sensor operates by oxygen quenching of induced room temperature phosphorescence. (p. 349)

8:30am **FAA3 New Fiber-Optic Sensor for Bile Reflux,** F. Baldini, R. Falciai, Annamaria Scheggi, *IROE-CNR, Italy;* P. Bechi, *U. Firenze, Italy.* A new method for the detection of enterogastric and nonacid gastroesophageal reflux, based on the absorption properties of bile, is developed. The fiber-optic sensor system is described and experimental results are reported. (p. 353)

8:45am **FAA4 Interferometric Optical Path Difference Measurement Using Sinusoidal Frequency Modulation of a Diode Laser,** D. J. Webb, R. M. Taylor, J. D. C. Jones, D. A. Jackson, *U. Kent, U.K.* We describe a technique applicable to interferometric systems illuminated by a laser diode, whereby the optical path difference is recovered by means of sinusoidal modulation of the laser emission frequency. (p. 357)

9:00am **FAA5 Wide Range Sensing of Liquid Refractive Index,** Elric W. Saaski, Gordon L. Mitchell, James C. Hartl, *Technology Dynamics, Inc.* The refractive index of fluids is measured utilizing an etched resonant cavity. The spectral reflectivity varies with index and is measured ratiometrically. Resolution of 0.0001 R.I. has been demonstrated over 1.0–2.0 R.I. (p. 361)

9:15am **FAA6 Remotely Sensing Molecular Spectra with Fluoride Fiber,** Pierre de Rochemont, Philip Levin, *SpecTran Corp.;* Allen Bonde, Jr., *Worcester Polytechnic Institute;* Anthony Boniface, *U. Massachusetts, Amherst;* Daniel Rapp, *USMC Development Center.* Heavy metal fluoride fibers, transmissive over infrared wavelengths between 2 and 4 μm, are used to remotely sense molecular spectra by Fourier methods. We report on results obtained using direct absorption and evanescent sensor probes. (p. 365)

9:30am **FAA7 Plastic-Clad Silica Fiber Chemical Sensor for Ammonia,** L. L. Blyler, Jr., J. A. Ferrara, J. B. MacChesney, *AT&T Bell Laboratories.* Dye-doped silicone-clad silica fibers have been fabricated and evaluated as sensors for ammonia. The evanescent field of light propagating in the core interacts with the dye, which responds to ammonia diffusing into the cladding. (p. 369)

9:45am **FAA8 Porous Fiber-Optic for a High Sensitivity Humidity Sensor,** Mahmoud R. Shahriari, George H. Sigel, Jr., Quan Zhou, *Rutgers U.* A porous fiber-optic sensor for humidity measurements has been developed. The device offers greatly increased sensitivity compared to evanescent or fluorescent fiber moisture sensors as well as excellent response time and full reversibility. (p. 373)

FRIDAY, JANUARY 29, 1988—*Cont.*

MEETING ROOM 5-7-9

8:00–10:00am
FBB, Rotation Sensors
Ramon P. De Paula, *Jet Propulsion Laboratory, Presider*

8:00am **FBB1 Fiber-Optic Gyroscope: Another Optical Revolution in the Inertia Business,** Herve C. Lefevre, *Photonetics S. A., France.* After more than ten years of active research, the fiber-optic gyroscope has entered its development stage. We discuss this strong competitor to well established technologies in such areas as navigation, guidance, and control. (p. 384) **(Invited paper)**

8:30am **FBB2 Interferometric Fiber-Optic Gyroscope Using a Novel 3 × 3 Integrated Optic Polarizer/Splitter,** William J. Minford, Gail A. Bogert, *AT&T Bell Laboratories;* Ramon DePaula, *Jet Propulsion Laboratory.* A Ti:$LiNbO_3$ 3×3 directional coupler acting as polarizer and 3-dB splitter with a phase modulator has been incorporated in an interferometric fiber-optic gyroscope. A random walk performance of $1 \times 10^{-2°}/\sqrt{h}$ has been demonstrated. (p. 385)

8:45am **FBB3 Scale Factor Accuracy and Stability in an Open Loop Fiber-Optic Gyroscope,** R. P. Moeller, W. K. Burns, N. J. Frigo, *U.S. Naval Research Laboratory.* The effects of optical coherence and feedback on the output of an open loop gyroscope are demonstrated. Scale factor stability of 32 ppm was achieved for a 12-day run, allowing for thermal and source drive current variations. (p. 393)

9:00am **FBB4 Drift Reduction in the Optical Heterodyne Fiber Gyro,** Kazuo Hotate, Shigeatsu Samukawa, *U. Tokyo, Japan;* Noboru Niwa, *Chiba Institute of Technology, Japan.* The optical heterodyne fiber gyro we proposed has been improved to reduce the zero-point drift. We now introduce the reference path into the optical system and a signal-processing scheme. The experiments show successful drift reduction. (p. 397)

9:15am **FBB5 Effect of Reflections on the Drift Characteristic of a Fiber-Optic Passive Ring-Resonator Gyro,** Masanobu Takahashi, Shuichi Tai, Kazuo Kyuma, Koichi Hamanaka, *Mitsubishi Electric Corp., Japan.* A fiber-optic passive ring-resonator gyroscope using an external-cavity laser diode is reported. A rotation rate as low as 3×10^{-5} rad/s was obtained by thoroughly reducing the effects of reflected light. (p. 401)

9:30am **FBB6 Drift of an Optical Passive Ring-Resonator Gyro Caused by the Faraday Effect,** K. Hotate, M. Murakami, *U. Tokyo, Japan.* A theoretical investigation is presented for the drift caused by the Faraday effect in an optical passive ring-resonator gyro. We show ways of suppressing this drift. (p. 405)

9:45am **FBB7 Evaluation of Polarization-Maintaining Fiber Resonator for Rotation Sensing,** G. A. Sanders, N. Demma, G. F. Rouse, R. B. Smith, *Honeywell Systems & Research Center.* Characteristics of a polarization-maintaining fiber resonator were evaluated for gyro applications. A closed-loop gyro was assembled using bulk optic components to obtain preliminary output-vs-rotation data. Noise and drift were also measured. (p. 409)

FRIDAY, JANUARY 29, 1988—*Cont.*

MEETING ROOM 2-4-6

10:30am–12:15pm
FCC, Strain, Pressure, and Flow Sensors
Kazuo Hotate, *University of Tokyo, Japan, Presider*

10:30am **FCC1 Optical Fiber Sensors for Composite Structures,** James R. Dunphy, Gerald Meltz, *United Technologies Research Center.* The compatibility of optical fiber sensors with composite materials enables one to install internal monitoring devices prior to curing. Two specific devices are discussed for potential use in curing process control, integrity assessment during storage, and in-service health monitoring. (p. 414) **(Invited paper)**

11:00am **FCC2 Elliptical Core Two-Mode Fiber Strain Gauge with Heterodyne Detection,** J. N. Blake, Q. Li, B. Y. Kim, *Stanford U.* We report two methods of heterodyning an elliptical core two-mode fiber strain gauge to obtain an rf output signal whose phase is linearly proportional to the axial strain applied to the fiber. (p. 416)

11:15am **FCC3 Development of a Structurally Imbedded Fiber-Optic Impace Damage Detection System for Composite Materials,** R. M. Measures, N. D. W. Glossop, J. Lymer, R. C. Tennyson, *U. Toronto, Canada.* As part of a program to develop a structurally imbedded fiber-optic impact damage detection system for composite material structures, we have undertaken a careful study of the sensitivity of the optical fibers to their orientation with respect to the adjacent plies and their depth in regard to the impact surface location. (p. 420)

11:30am **FCC4 High-Sensitivity Fabry-Perot Optical Fiber Sensor to Measure Mechanical Force,** F. Maystre, P. Gannage, R. Dandliker, *U. Neuchatel, Switzerland.* A high-resolution polarimetric Fabry-Perot sensor concept with a diode laser source and heterodyne phase detection is presented. It has been successfully used to measure static force by induced birefringence with ten times increased sensitivity. (p. 424)

11:45am **FCC5 Optically Excited Micromechanical Resonator Pressure Sensor,** K. E. B. Thornton, D. Uttamchandani, Brian Culshaw, *U. Strathclyde, U.K.* Micromechanical silicon resonators have been optically excited into transverse vibrations. An optical interferometer was used to detect their vibration frequency. We report the pressure dependence of resonant frequency of this miniature resonant optical sensor. (p. 433)

12:00m **FCC6 Fiber-Optic Flow Sensor,** Michael H. Ikeda, Mei H. Sun, *Luxtron Corp.;* Stephen R. Phillips, *Alamo Instruments.* The development of a fiber-optic analog to the hot film anemometer is summarized. The infrared heated optical sensor loses its thermal energy to the flowing medium. The heat loss is correlated to the local flow rate. (p. 437)

FRIDAY, JANUARY 29, 1988—*Cont.*

MEETING ROOM 5-7-9

10:30am–12:00m
FDD, Sources for Sensors
Scott Rashleigh, *Australian Optical Fiber Research Party Ltd., Australia, Presider*

10:30am **FDD1 Rare Earth Fibers,** Elias Snitzer, *Polaroid Corporation.* The early work on soft glass single-mode fiber lasers for narrow band amplifiers and low noise detectors is reviewed. The active ions considered include neodymium and ytterbium and the configurations are for oscillators, amplifiers, and superluminescent sources. (p. 448) **(Invited paper)**

11:00am **FDD2 High Power GaAlAs Superluminescent Diode for Fiber Sensor Applications,** Norman S. K. Kwong, Kam Y. Lau, Nadav Bar-Chaim, Israel Ury, Kevin Lee, *Ortel Corp.* Superluminescent diodes with high output power (22.5 mW) at low injection current (100 mA) and with a stable transverse mode are reported. The spectral modulation depth is below 25% over the entire emission spectral bandwidth of 20 nm. (p. 451)

11:15am **FDD3 Chirped Semiconductor Laser as an Alternative to the SLD in a Fiber Gyro.** L. Hergenroeder, S. P. Smith, S. Ezekiel, *MIT Research Laboratory of Electronics.* A chirped semiconductor laser is used instead of a superluminescent diode to reduce the effects of backscattering and the optical Kerr effect in a fiber interferometer gyroscope. (p. 455)

11:30am **FDD4 Optical Feedback Effects on Superluminescent Diodes,** R. O. Miles, *Sachs/Freeman Associates;* W. K. Burns, R. P. Moeller, A. Dandridge, *U.S. Naval Research Laboratory.* We report on the effect of optical feedback on fiber pigtailed superluminescent diode sources. We present and discuss the effects of Rayleigh scattering and direct feedback on the spectral output of the superluminescent diode. (p. 458)

11:45am **FDD5 Superfluorescent Single-Mode Nd:Fiber Source at 1060 nm,** K. Liu, M. Digonnet, K. Fesler, B. Y. Kim, H. J. Shaw, *Stanford U.* A broadband source in single-mode Nd-doped fiber shows the high output power (10 mW) and spectral width of 17 nm FWHM desirable for fiber-optic gyroscopes. Wavelength stability as a function of temperature and pump parameters is discussed. (p. 462)

FRIDAY, JANUARY 29, 1988—*Cont.*

MEETING ROOM 2-4-6

1:30–3:00pm
FEE, Polarization-Maintaining Components
Eric Udd, *McDonnell Douglas Astronautics Company, Presider*

1:30pm **FEE1 Fiber Devices for Fiber Sensors,** Juichi Noda, Itaru Yokohama, *NTT Opto-electronics Laboratories, Japan.* Fiber devices such as couplers, polarizers, depolarizers, polarization controll ers, filters, nonreciprocal devices, and lasers are reviewed. Current results on polarization independent/dependent fiber couplers essential for fiber sensors are mainly discussed. (p. 468) **(Invited paper)**

2:00pm **FEE2 High-Extinction-Ratio and Low-Loss Single-Mode Single-Polarization Optical Fiber,** K. Himeno, Y. Kikuchi, N. Kawakami, O. Fukuda, K. Inada, *Fujikura, Ltd., Japan.* A highly birefringent fiber for single-mode single-polarization operation has been fabricated. The fiber with modal birefringence of above 8×10^{-4} exhibits low loss of 2.7 dB/km and an extinction ratio of 47 dB in short and/or long length. (p. 472)

2:15pm **FEE3 High Performance Polarizers and Sensing Coils with Elliptical Jacket Type Single-Polarization Fibers,** Y. Takuma, H. Kajioka, K. Yamada, *Hitachi Cable, Ltd., Japan.* In-line polarizers using elliptical jacket-type single polarization fibers with an extinction ratio of < -40 dB and an insertion loss of < 0.5 dB have been developed. A sensing coil with an extinction ratio of -45 dB/0.5 km and no excess loss has also been developed. (p. 476)

2:30pm **FEE4 Ultralow Crosstalk Polarization-Maintaining Optical Fiber Coupler,** T. Arikawa, F. Suzuki, Y. Kikuchi, O. Fukuda, K. Inada, *Fujikura, Ltd., Japan.* The experimental evaluation of polarization-maintaining optical fiber couplers fabricated by a polishing technique is described. The couplers show polarization crosstalk of less than -30 dB and small wavelength dependence of the splitting ratio. (p. 480)

2:45pm **FEE5 Polarization-Controlled Fiber-Optic Recirculating Delay Line Filter and its Associated Phase-Induced Intensity Noise,** Ady Arie, Moshe Tur, *Tel Aviv U., Israel.* By inserting a linear polarizer into a fiber-optic recirculating delay line, the system becomes high polarization dependent. The transfer function and the phase-induced intensity noise for different polarization inputs are studied. (p. 484)

3:00–4:00pm
Postdeadline Paper Session
Frederick J. Leonberger, *United Technologies Research Center, Presider*

FRIDAY, JANUARY 29, 1988—*Cont.*

MEETING ROOM 5-7-9

1:30–2:45pm
FFF, Temperature Sensors
Gordon Mitchell, *Metricor, Inc., Presider*

1:30pm **FFF1 Phase Measurement Based Ruby Fluorescence Fiber-Optic Temperature Sensor,** K. T. V. Grattan, R. K. Selli, A. W. Palmer, *City U., U.K.* A fluorescence-based point temperature sensor using ruby excited by sinusoidally modulated light from a LED is described. Phase-lag changes with temperature of the received sinusoidally modulated fluorescence are determined using simple, yet sophisticated microprocessor-based electronics. (p. 490)

1:45pm **FFF2 Simultaneous Measurement of Temperature and Pressure Variations with a Single-Mode Fiber,** D. Chardon, *Institut d'Optique, France;* S. J. Huard, *ENSPM, France.* Pressure and temperature action on a single-mode fiber wrapped around a cylinder affects both birefringence and mean phase. Simultaneous measurement of these quantities unambiguously provides the values of the two physical parameters. (p. 495)

2:00pm **FFF3 Noncontact Temperature Measurement with a Zirconium Fluoride Glass Fiber,** Serge Mordon, Elizabeth Zoude, Jean Marc Brunetaud, *INSERM, France.* Infrared signal transmission (1-5 μm) through zirconium fluoride glass fibers is used for noncontact temperature measurement. Excellent accuracy can be obtained for temperatures as low as 60°C and with a time constant equal to 1 ms. (p. 502)

2:15pm **FFF4 Simple Fiber-Optic Fabry-Perot Temperature Sensor,** L. Schultheis, *Brown Boveri Research Center, Switzerland.* A simple fiber-optic temperature sensor based on the thermal shift of the Fabry-Perot interferences of a semiconductor etalon is investigated. Different semiconductors as well as various fabrication techniques are used and the implications to the sensitivity of the sensor are discussed. (p. 506)

2:30pm **FFF5 Optical Fiber Thermal Conductivity Sensor,** B. J. White, J. P. Davis, L. C. Bobb, *U.S. Naval Air Development Center;* D. C. Larson, *Drexel U.* A Mach-Zehnder interferometer was used to determine the thermal conductivities of liquids by measuring the temperature vs time of a resistively heated section of conductively coated fiber in one arm of the interferometer. (p. 510)

THURSDAY, JANUARY 28, 1988

MEETING ROOM 2-4-6

1:30 PM–3:15 PM

ThDD1–5

INTEGRATED OPTIC DEVICES

Masamitsu Haruna, Osaka University, Japan, ***Presider***

Passive Integrated Optic Devices

R.Th.Kersten
SCHOTT Glassworks, Mainz

W.Roß
Integrated Optic Technology, Heidelberg

Abstract

Fiber and integrated optic sensors are gaining importance. For the new generation of integrated and fiber optic sensors the micro- and fiber-optic elements will be replaced by integrated optics. IO-technologies, applications of optical sensors and IO-elements will be discussed. Specific examples will be presented.

Summary

Optical sensors are of great interest because of their immunity to strong electric fields, their high sensitivity and (sometimes) selectivity. Compared to conventional optical sensors, fiber and integrated optic sensors have the advantage to be very small. A fiber tip being the sensor element can reach any place, i.e. even inside the human body.

Fiber optic sensors need certain optics for coupling light in and out and for signal processing. Today this is accomplished by using micro- or fiber-optic devices. In future, integrated optics will be preferred because of high reliability, small size, mechanical stability and last but not least an expected low price. Also the integration of several functions onto a single chip will be an asset.

Technologies for fabrication of passive integrated optical circuits emphasizing ion exchange in glass will be discussed. It will be shown, that it is necessary to use special glasses to improve the properties (e.g. waveguide losses) as well as the fabrication process (e.g. for improved reproducibility) of ion exchanged waveguides. Differences with respect to multimode and single-mode waveguide technology will be addressed.

A variety of integrated optic applications will be demonstrated by using examples such as fiber gyros, resonant IO-gyros, and evanescent field sensors. The advantages and disadvantages as well as existing problems will be discussed.

First approaches to integrated optic structures for the use in/for optical sensors will be shown. Their properties and applications will be discussed. Examples of different types of IO-waveguides for/as optical sensors will be given.

ACTIVE INTEGRATED OPTICS FOR SENSORS

MASAMITSU HARUNA and HIROSHI NISHIHARA

Osaka University, Faculty of Engineering,
Department of Electronics
2-1 Yamada-Oka, Suita, Osaka 565, Japan

Recently, there is considerable interest in integration of some waveguide components on a common substrate to achieve a compact and rugged fiber-optic sensor [1]. Among a variety of fiber-optic sensors, the heterodyne-detection optics is very often used for highly accurate measurement of velocity, displacement and position. This particular optics requires active optical components such as phase modulators, frequency shifters, wavelength plates, and so on. Such active components can be replaced by $LiNbO_3$ waveguide devices, which have essentially much higher-speed capability with extremely lower drive voltage compared to the bulk-optic counterparts. $LiNbO_3$ is thus a promising waveguide material for integrated-optic sensors. This paper presents our recent research activities on optical integration of the heterodyne optics, especially for a fiber laser Doppler velocimeter (fiber LDV).

A typical heterodyne optics for velocity measurement which consists of bulk-optic components is shown in Fig. 1, where a polarization-maintaining fiber (PM fiber) is used for picking up a Doppler-shifted signal. In full integration of such a fiber LDV, the Bragg cell, half-wave plate and polarization beam splitter can be replaced by the waveguide frequency shifter, TE/TM mode converter and mode splitter, respectively. The Mach-Zehnder waveguide interferometer is also constituted by an aluminum-film mirror on the waveguide end and Y-junction waveguides having the function of power division/combination. The whole heterodyne optics for velocity measurement can thus be integrated on a $LiNbO_3$ substrate, as shown in Fig. 2 [2]. Since a Z-propagating $LiNbO_3$ is used as a substrate to avoid the optical damage, the total length of the optical IC then becomes 32 mm, while the guide width is only 3.5 μm for single-mode propagation at the 0.633- μm wavelength. Such a large-area waveguide patterning is possible within 0.2- μm accuracy by the computer-controlled laser-beam lithographic system, developed recently in our laboratory, which can be widely used for the practical optical-IC patterning [3,4]. The optical IC was then pigtailed with a PM fiber having 4- μm core diameter in the same manner as reported by Cameron [5]. The fiber-to-waveguide coupling efficiency of nearly 70 % was obtained. Using the fabricated optical IC, the velocity of a moving mirror was successfully measured with the signal-to-noise ratio (S/N ratio) of 25 dB [6].

Presently, our effort is directed at development of a hybrid optical IC with a LD used as a light source. Such a hybrid optical IC should be designed so that the optical-path length difference between the reference and signal arms is shorter than the coherent length of the LD. The detailed experimental result will be presented at the Conference.

Next, we propose a new optical IC for 2-D velocity measurement designed and fabricated based on the integrated-optic fiber LDV, as described above. The schematic view is shown in Fig. 3, in which a waveguide switch was integrated in addition to the frequency shifter. Two single-mode fibers were butt-coupled into the outputs of the so-called balanced-bridge switch with 10-mm long electrodes. This polarization-independent switch exhibited the extinction ratio of >20 dB with the half-wave voltage of 8 V. The fiber-to-waveguide coupling loss was also nearly 3 dB after permanent fixing with UV-light-cured epoxy. Using the fabricated optical IC, both velocity components v_x and v_y of a moving right-angle-corner mirror were measured. The switch was driven by a pulse voltage with repetition rate of 1.5 MHz which was much higher than the shifted-frequency f_R = 300 KHz of the serrodyne frequency shifter. Fig. 4 shows the resulting frequency spectra obtained without and with the switching gate driven in synchronization with the pulse voltage of 1.5 MHz. The time-division 2-D velocity measurement has thus been demonstrated with the S/N ratio of >20 dB.

In addition, an optical IC of the two-frequency polarization interferometer, used for highly accurate position sensing, was also fabricated and tested [3]. This optical IC includes a two-frequency shifter with mutually orthogonal polarizations which allows us to use a linear-polarized laser as a light source instead of a Zeeman laser, the TE/TM mode converter, mode splitter and some 3-dB directional couplers. Although the optical IC is rather sophisticated and as long as 45 mm, the waveguide patterning is performed relatively easily by our laser-beam lithographic system. It was confirmed to date that, using the fabricated optical IC, the velocity of a moving object was measured with the S/N ratio of nearly 20 dB. We have continued to examine the measurement characteristic of phase shift of the signal light corresponding to displacement of the object [7].

We have presented our recent progress in optical integration of the heterodyne optics used for fiber-optic sensors. The integrated-optic technology based on $LiNbO_3$ is actually useful not only for miniaturization of fiber-optic sensors but also for development of new sensing devices with high-speed signal processing capability. For instance, the optical IC for 2-D velocity measurement proposed here indicates the possibility of high-speed time-division multi-point sensing.

References

1. for example, H. Toda, M. Haruna and H. Nishihara, Electron. Lett. **22**, p.982 (1986).
2. H. Toda, M. Haruna and H. Nishihara, OFS'86, Paper 4.7, Tokyo, Oct 1986.
3. M. Haruna and H. Nishihara, OFC/IOOC'87, TuH6, Reno, Nevada, Jan. 1987.
4. M. Haruna, S. Yoshida, H. Toda and H. Nishihara: to be published in Appl. Opt. **26**, Nov. 1987.

5. K.H. Cameron, Electron. Lett. **20**, p.974 (1984).
6. H. Toda, M. Haruna and H. Nishihara, IEEE J. Lightwave Tech. **LT-5**, July 1987.
7. The detailed description will appear elsewhere.

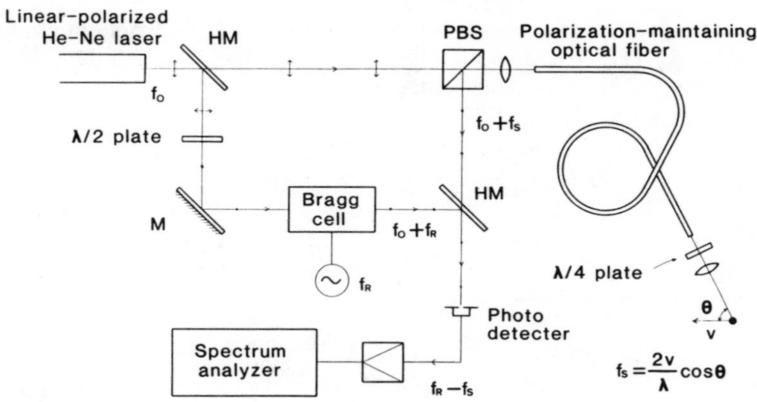

Fig. 1 A heterodyne optics for the fiber LDV.

Y-junction waveguide (1/50rad) Directional waveguide coupler

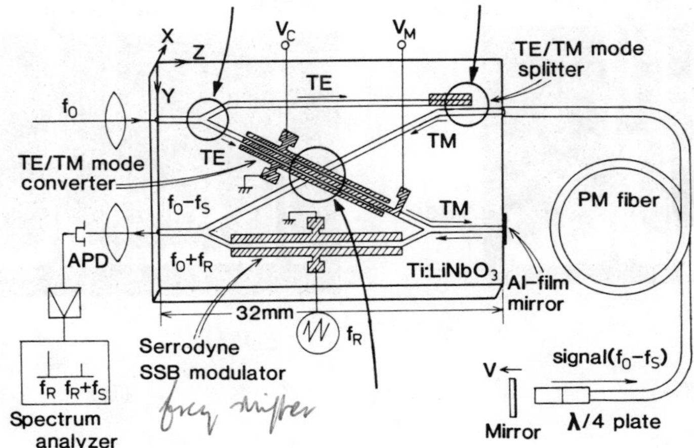

Crossed waveguide (1/25rad)

Fig. 2 Integrated-optic fiber LDV, and microscope photographs of Ti-diffused waveguides defined by the laser-beam lithographic system.

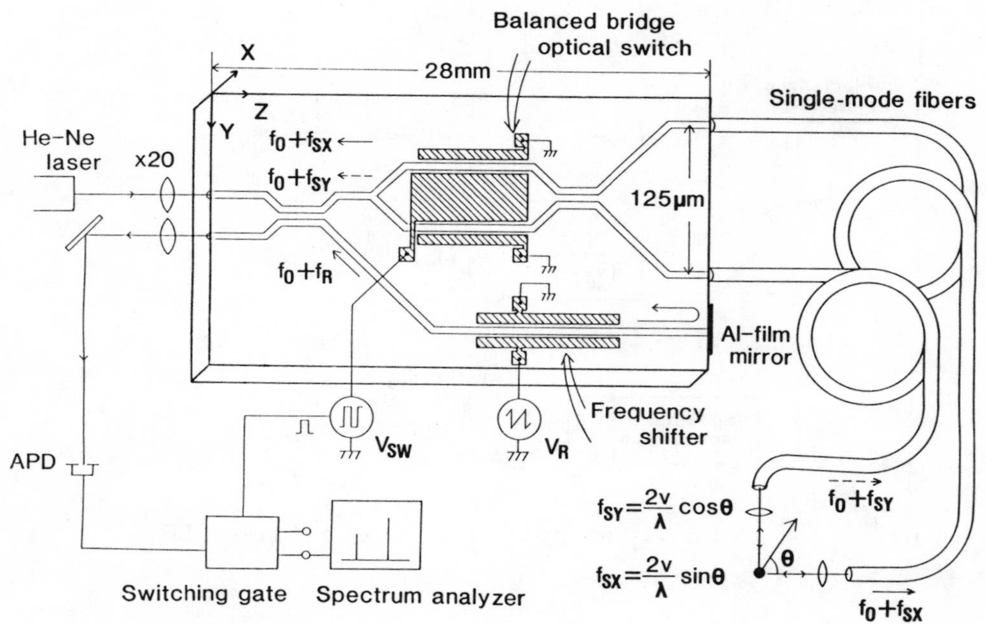

Fig. 3 Optical IC for time-division 2-D velocity measurement.

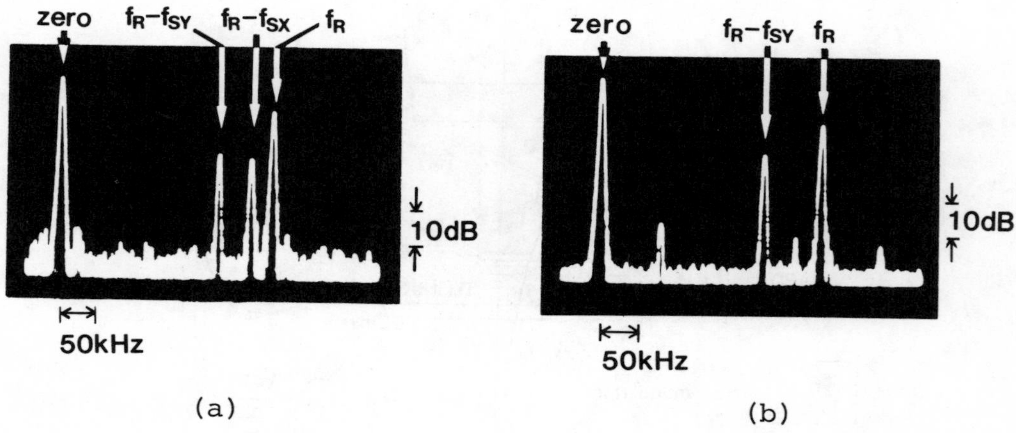

(a) (b)

Fig. 4 The resulting frequency spectra obtained (a) without and (b) with the switching gate. The frequency shift f_R of the reference light is 300KHz. Doppler-shifted frequences f_{sx} and f_{sy} correspond to velocity components v_x and v_y, respectively.

Ti-$LiNbO_3$ Waveguide Serrodyne Modulator With Ultrahigh Sideband-Suppression for Fiber Optic Gyroscope

H. Hung, C. Laskoskie, T. El-Wailly, and C. L. Chang

Honeywell Inc.

System and Research Center

P.O. Box 21111

Phoenix, Arizona 85036

(602) 869-6344

SUMMARY

Serrodyne modulation using an electro-optic (E-O) phase modulator in a Ti-indiffused $LiNbO_3$ waveguide for optical frequency shifting is being developed for a variety of applications.[1] Sideband suppression better than 40 dB at a moderate shifting frequency (40 KHz) has been reported.[2] Inertial navigation, phase-nulling rotation sensors utilizing Sagnac interferometers, however, require extremely high sideband suppression and shifting frequencies of up to several hundred kilohertz in order to achieve scale factor linearity over a wide dynamic range.[3-4] In this paper, we report a serrodyne modulator with an overall sideband suppression of better than 50 dB at frequency shifts of up to 200 KHz.

To realize serrodyne modulation using a Ti-$LiNbO_3$ waveguide phase modulator, we have built a signal generator which produces a sawtooth signal with a fly-back time of 5 ns and repetition rates from 6 KHz up to 200 KHz. The phase modulator has flat frequency response from DC to 450 MHz and requires a 5V P-P signal to obtain a 2π radian phase shift. Figure 1 illustrates a schematic of the Mach-Zehnder interferometer operated at 830 nm wavelength used for device testing. One arm of the interferometer contains the phase modulator under test while an acousto-optic Bragg cell providing a 35 MHz shift to the first-order diffracted beam is placed within the second arm. The two beams, after traversing their respective modulators, are recombined at the output coupler and fed into a photodetector. This simple heterodyne detection allows us to measure the optical output spectrum. Figure 2 shows the typical experimental results. As can be seen, all the spurious sideband levels are 55 dB below the desired signal. Preliminary theoretical results indicate that the sideband suppression can be further improved by utilizing a phase modulator with a broader bandwidth. Furthermore, the fly-back time is found to be equally important to the sideband suppression in our device. Work on high-speed electronics for producing faster fly-back time and extending the operation frequency to a viable range is in progress.

Theoretical and experimental data for sideband suppression S, versus peak-to-peak serrodyne phase shift ϕ, using a 2 KHz ramp with 2 μsec flyback, are shown in Figure 3. It is noted that near the optimum phase shift ϕ_0, sidebands suppression becomes increasingly sensitive to peak-to-peak phase shift errors. This is theoretically analyzed according to the relationship, $\Delta S = -20 \text{ Log } [(\phi - 2\pi)/(\phi_0 - 2\pi)]$. For the experimental data, the maximum obtainable suppression was 48 dB. If 40 dB is taken as a specified suppression value, a required peak phase stability of 0.6% is calculated. This is found in good agreement with the experimental value, 0.5%.

In summary, we have confirmed that high sideband suppression for a serrodyne modulator using a broadband Ti-$LiNbO_3$ waveguide modulator can be obtained when a linear ramp signal with fast fly-back time is applied to the device. Furthermore, it was shown that phase shift stability becomes very critical for the attainment of ultrahigh sideband suppression.

REFERENCES

1. K. K. Wong, S. Wright, and R. M. DeLaRue, "Topical Meeting on Integrated and Guided-Wave Optics", Optical Society of America, Washington, D. C., paper WA5, 1986.

2. L. M. Johnson and C. H. Cox, III, "Topical Meeting on Integrated and Guided-Wave Optics, Optical Society of America", Washington, D. C., paper WBB4, 1986.

3. R. F. Cahill and E. Udd, Opt. Lett., 4, 93, 1979.

4. J. L. Davis and S. Ezekiel, Opt. Lett., 6, 505, 1981.

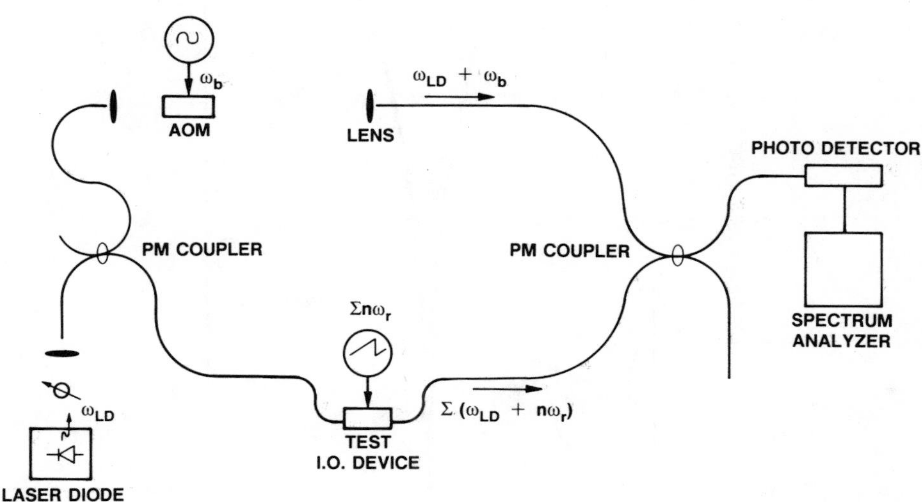

Fig. 1 Schematic of Mach-Zehnder interferometer used for serrodyne modulator evaluation.

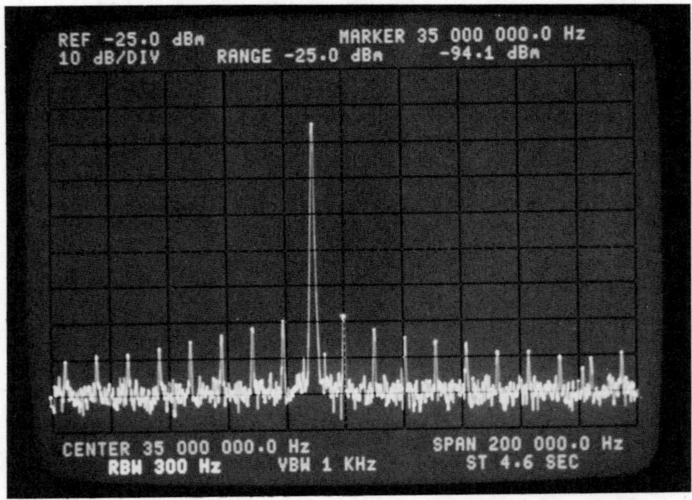

Fig. 2 Overall sideband suppression measurement for Ti-$LiNbO_3$ serrodyne modulator showing greater than 55 dB suppression at shifting frequency of 10 kHz.

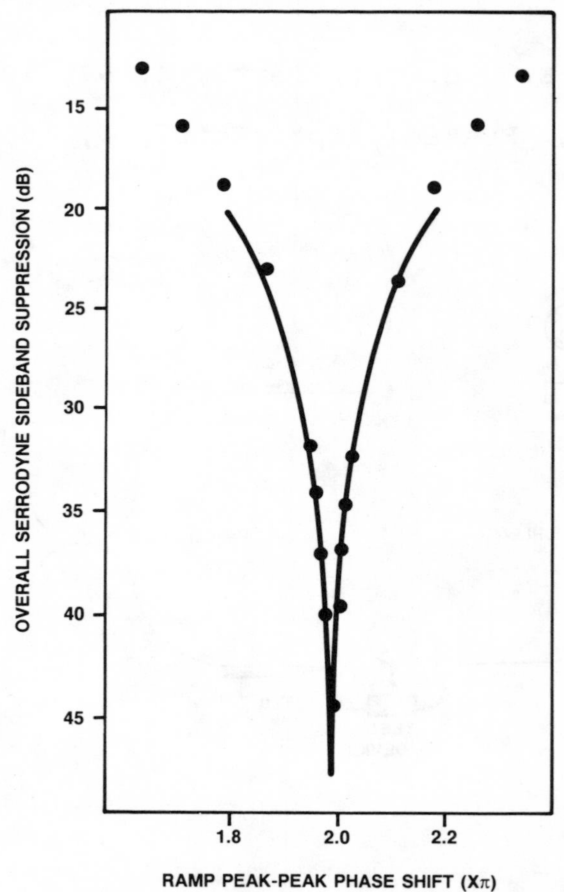

Fig. 3 Theoretical and experimental results for sideband suppression versus peak-to-peak serrodyne phase shift.

Fiber Optic Interferometer with Digital Heterodyne Detection Using Lithiumniobate Devices

D. Eberhard*, E. Voges

Universität Dortmund, Lehrstuhl für Hochfrequenztechnik
D - 4600 Dortmund 50 , Federal Republic of Germany

Fiber optic interferometers with single mode fibers provide high sensitivity for sensing e.g. temperature, sound, electric and magnetic fields. The inclusion of integrated optic components into the interferometer implements important functions such as splitting, recombining, phase compensation, polarization control in a compact way. Above all, integrated optic devices allow efficient heterodyne detection schemes with a linear conversion of the sensor phase Θ and the sensor transmission H into corresponding electrical signals. Electro-optic $LiNbO_3$ devices are particularly attractive for these purposes. Here, we report on fiber optic Michelson and Mach-Zehnder interferometers which are built-up with polarization maintaining fibers and integrated optic $LiNbO_3$ devices. A digital heterodyne detection with an high dynamic range (up to 60 dB) is accomplished by a proper digital phase modulation. A basic configuration is depicted in Fig.1. Electro-optic phase modulators are driven with digital waveforms through a microcomputer that also samples the diode current. Due to the high linearity of $LiNbO_3$ phase modulators, a staircase phase modulation $\phi(t)$ of phase modulator PM 1 /1/ in analogy to the analogue serrodyne /2/ or sawtooth /3/ modulation, proves to be efficient. A digital fourier transform (DFT) of the sampled diode current yields the sensor transmission H and the sensor phase Θ. The proper peak phase deviation $\phi = 2\pi$ within a period τ is controlled via the zero of the second harmonic of the diode current. For 10-bit AD-converters the quantization of the diode current samples leads to maximum phase error $\Theta_Q = 0.065^o$ for the minimum number N = 3 of samples. For N = 100 samples per period we have $\Theta_Q = 0.005^o$. The phase error introduced by nonexact values $\phi = 2\pi$ of the peak phase deviation or by a parasitic amplitude modulation can be decreased below the above values by a second phase modulator. It is switched from period to period by 0° / 90° / 180° / 270°. This error reduction by cascaded phase modulators is similar to that in microwave network analysis employing binary phase modulation /4/. Fig. 3 and Fig. 4 show X the configurations of realized Michelson and Mach - Zehnder interferometers. The $LiNbO_3$ devices were fabricated on X - cut or Y - cut $LiNbO_3$ by conventional titanium indiffusion. For beam splitting BOA - Couplers /5/ are utilized.

* now at the Fraunhofer - Institut für Physikalische Meßtechnik, D - 7800 Freiburg

The phase modulators with 6 mm length and 9 μm electrode gap had U_π = 6 V switching voltage.
Polarization maintaining fibers (YORK HB 800/2) were carefully aligned, and attached one-by-one with UV-curing epoxy. The insertion loss of the fiber coupled devices typically amounted to 4-6 dB. For a strong (> 50 dB) suppression of the cross polarization, thin film polarizers (metal: Au, buffer: 30 nm Si_3N_4) were integrated as shown in Fig. 4.
The endfaces of the sensor and reference fibers in the Michelson interferometer of Fig. 3 were coated with evaporated Al-mirrors. The high back-reflection of the Michelson arrangement severely degraded the coherence of the laser diode if this was efficiently coupled to the input fiber. The Mach-Zehnder arrangement of Fig. 4 is free from strong back reflections.
One test result is shown in Fig. 5. A length of 3.5 cm of the sensor fiber in Fig. 4 was inserted into a Peltier element with ±1 mK temperature control. The measured time variation $\pm 0.2^\circ$ of the sensor phase Θ in Fig. 5 is due to this ±1 mK temperature control. The measurement accuracy of the digital heterodyne detection was better than 0.06° using N = 4 sampling points per period.
These results demonstrate the accurate phase and amplitude detection in fiber optic sensors by employing $LiNbO_3$ integrated optic devices and low-cost digital electronics.

References

/1/ R.J.King, Microwave homodyne systems, Peter Peregrinus Ltd., Stevenage, Great Britain 1978

/2/ E.Voges, O.Ostwald, B.Schiek and A.Neyer, IEEE J. Quantum Electron. QE-18, 124-129 (1982)

/3/ D.Eberhard, E.Voges, Proceedings of the 2nd International Conference on Optical Fiber Sensors, Stuttgart, 1984, 381-384

/4/ U.Gärtner, B.Schiek, IEEE Trans. Microwave Theory Techn. MTT-34, 902-906 (1986)

/5/ M.Papuchon, A.Roy and D.Ostrowsky, Appl.Phys.Lett. 13, 266-267 (1977)

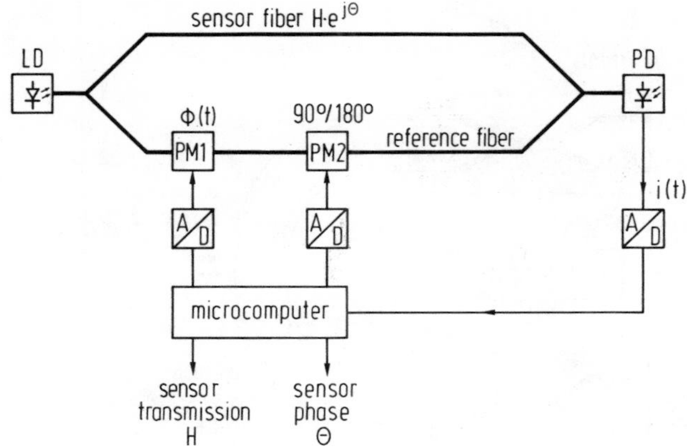

Fig. 1
Fiber-optic Mach-Zehnder interferometer with a sensor fiber of amplitude transmission H and phase shift Θ to be measured and heterodyne detection by digital phase modulation (LD: laser diode, PD: photo diode, PM: phase modulator)

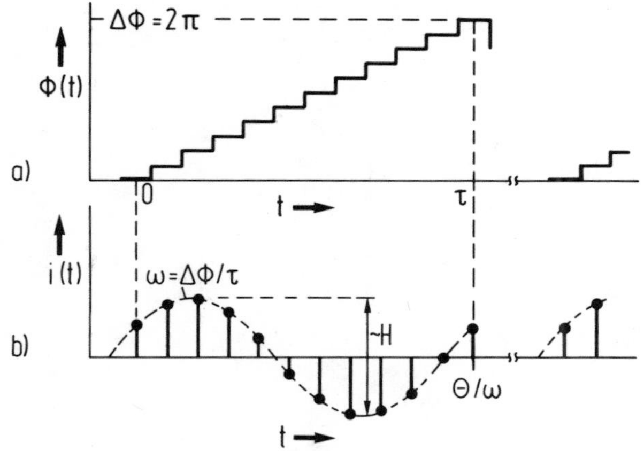

Fig. 2
a) Staircase phasemodulation $\phi(t)$ with $N \geq 3$ steps and $\phi = 2\pi$ phase shift within a periode τ
b) sampled diode current with measured values Θ and H

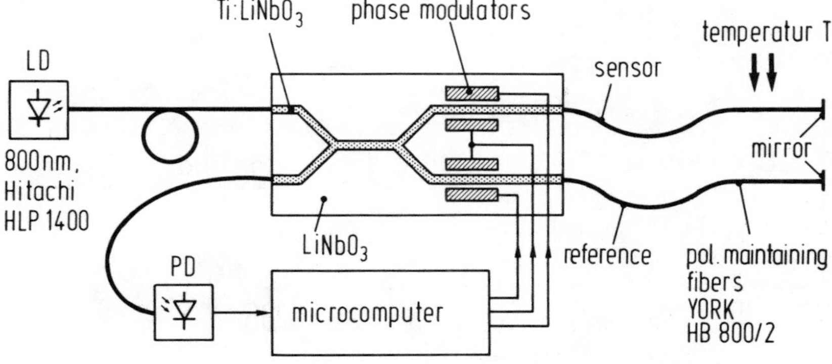

Fig. 3
Michelson - Interferometer sensor with $LiNbO_3$-chip for splitting and phase modulation

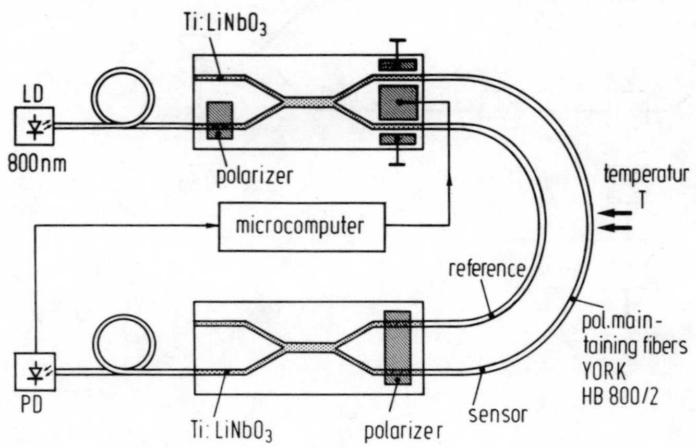

Fig. 4
Fiber-optic Mach-Zehnder interferometer sensor with $LiNbO_3$ integrated optic components

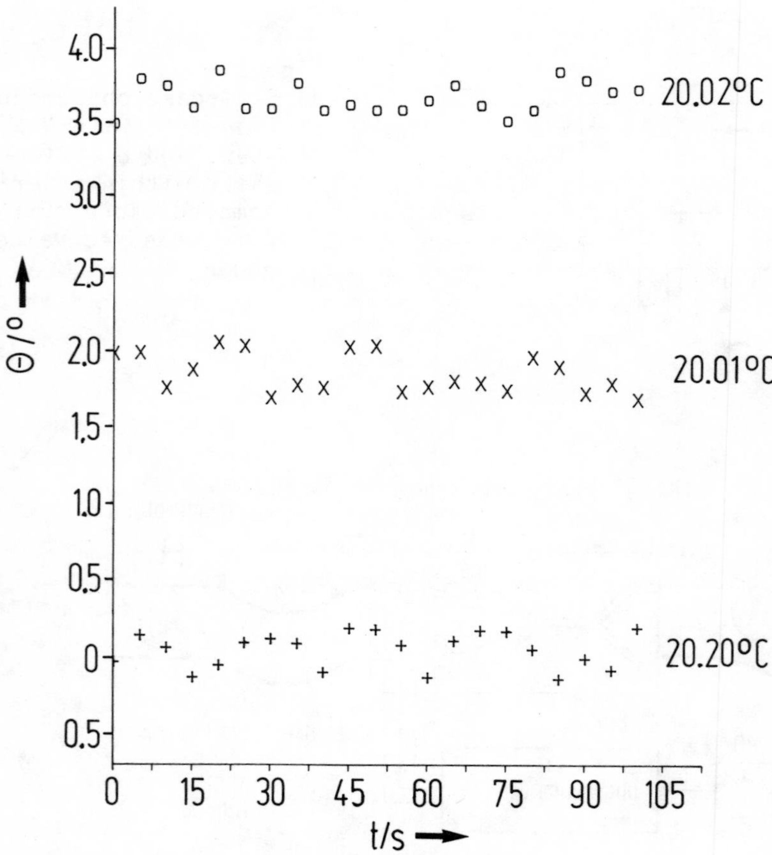

Fig. 5
Measured optical phase Θ in dependence on time t for three temperatures T applied to part of the sensor fiber

High Quality Integrated Optical Polarizers in $LiNbO_3$

P.G. Suchoski, T.K. Findakly, and F.J. Leonberger
United Technologies Research Center
Silver Lane
E. Hartford, CT 06108

Introduction

Ti: $LiNbO_3$ integrated optical circuits are playing an increasingly important role in fiber optic sensor technology. To obtain acceptable performance in these systems, the power in the $LiNbO_3$ circuit must be linearly polarized and confined to the extraordinary mode. However, due to fiber misalignment and depolarization in the Ti: $LiNbO_3$ waveguides, polarization crosstalk levels are typically in the -25 to -35 dB range [1]. This problem can only be remedied by incorporating an integrated optical polarizer into the circuit. In this work, we describe a proton-exchanged polarizer operating at 0.8-0.85 μm in x-cut $LiNbO_3$ which provides 55-60 dB polarization extinction with very low (0.2 dB) excess loss. This represents a 15-20 dB improvement over the best integrated optical polarizer results reported to date in $LiNbO_3$ [2,3].

Fabrication and Characterization

The proton-exchanged polarizer [4] consists of Ti-diffused channel waveguides with short discontinuities which are filled with proton-exchanged waveguides as shown in Fig. 1. In the proton-exchanged waveguides $\Delta n_e > 0$ while $\Delta n_o < 0$. Thus, the extraordinary mode is guided along the entire length of the structure and experiences minimal insertion loss. The ordinary mode, on the other hand, is not guided in the proton-exchanged region and is radiated into the substrate. If the proton-exchanged region is sufficiently long, only a small portion of the radiated power is captured by the output Ti-diffused waveguide or fiber, resulting in high loss for the ordinary mode.

In this work, single mode waveguides with short discontinuities were initially fabricated on 3-cm-long x-cut $LiNbO_3$ crystals using Ti in-diffusion. The Ti thickness (40 nm) and strip width (5 μm) were chosen to yield waveguides with a single well-confined TE mode at 0.82 μm. An Al mask was subsequently deposited on the substrate and patterned so as to leave the crystal exposed where proton exchange was to take place. The proton-exchanged waveguides were then formed by exchanging in pure benzoic acid and annealing. As shown in Fig. 1, three polarizer geometries were considered: one- and two-section polarizers located in the middle of the sample and onesection polarizers located at the output of the sample. In all cases, L ranged from 0.25 to 4 mm in 0.25 mm increments. Ti-diffused waveguides without polarizers were also included on the samples.

The samples were characterized at 0.82 μm. Power was coupled into and out of the waveguides using 20x, strain-free objectives. A Glann-Thompson polarizer was used to polarize the input light while a second polarizer was used to analyze the waveguide output. The polarization extinction was determined by exciting the waveguide with the TE mode and then dividing the output TM mode power by the output TE mode power. The excess loss was determined by comparing the TE mode loss in waveguides with and without polarizers. The dynamic range of the PER measurements was 63 dB and was limited by the Glann-Thompson polarizers. The insertion loss measurements were repeatable to within 0.1 dB.

Polarizer Performance

Excess insertion loss in the proton-exchanged polarizer is a result of two factors: propagation loss in the proton-exchanged waveguide and mismatch between the optical mode size in the Ti-diffused waveguide and the proton-exchanged waveguide. The Ti-diffused waveguides used in our integrated optical circuits have a 1/e intensity full width of 5.8 μm, a full depth of 3.3 μm, and a propagation loss of 0.25-0.3 dB/cm. The two-step proton exchange process developed for this work yields waveguides with a 1/e intensity full width of 4.9 μm, a full depth of 3.2 μm, and a propagation loss of 0.2-0.3 dB/cm. Evaluation of the overlap integral indicates the proton-exchanged waveguides should exhibit 0.2 dB mode mismatch with the Ti-diffused waveguides per interface. The measured excess insertion loss is 0.5 dB for single section polarizers in the middle of the sample (2 interfaces), 1.0 dB for two-section polarizers (4 interfaces), and 0.2 dB for single-section polarizers at the output of the sample (1 interface).

The PER of the proton-exchanged polarizer is a function of both the length and the location of the proton-exchanged waveguide. To determine how the PER varies with length, the samples are excited with the TM mode and the output polarizer is adjusted to pass the TM mode power. The TM mode extinction is defined as the measured TM mode power from a waveguide with a polarizer divided by the measured TM mode power from a waveguide without a polarizer. By adopting this approach, it is possible to accurately estimate the polarizer's extinction properties without being strongly influenced by depolarization in the Ti-diffused waveguides. A plot of the TM mode extinction versus the total polarizer length is presented in Fig. 2. Although this figure shows a measurement limit of 50 dB, we have recently improved this limit to 63 dB by using higher-quality objectives. Our results indicate that better than 60 dB extinction can be obtained for the TM mode by using one 2.5-mm-long proton-exchanged section or two 0.75-mm-long (total length L = 1.5 mm) proton-exchanged sections separated by a 5-mm-long Ti-diffused waveguide.

In actual practice, however, the input polarization would be adjusted such that most of the power is coupled into the TE (extraordinary) mode. When the $LiNbO_3$ waveguides are excited in this manner, the PER is limited to 32-35 dB (rather than the 60 dB we would expect from TM mode extinction measurements) due to depolarization in the output Ti-diffused waveguides. We have seen similar levels of depolarization in all Ti: $LiNbO_3$ waveguides fabricated in our laboratory and suspect this may be an inherent limitation which arises from waveguide imperfections introduced during fabrication [1]. This matter will be discussed in more detail at the conference.

The depolarization problem can be overcome by placing the proton-exchanged section at the output of the sample as shown in Fig. 1. For this geometry, we are able to obtain between 55 and 60 dB PER for TE mode excitation for a proton-exchanged output section 2.5 mm long or greater. The excess loss for this configuration is 0.2 dB. These values represent the highest PER and the lowest excess loss of any integrated optical polarizer reported to date in $LiNbO_3$. To further demonstrate the effect of depolarization, we rotated the sample so that the proton-exchanged section was at the input and re-characterized the sample. The measured PER decreased to 32-35 dB, indicating that our earlier PER measurements were indeed limited by depolarization in the Ti-diffused waveguides rather than limitations in the proton-exchanged polarizers.

Conclusion

In conclusion, we have presented a proton-exchanged polarizer in x-cut $LiNbO_3$ which exhibits a PER of 55-60 dB and an excess insertion loss of 0.2 dB at 0.82 μm. This was accomplished by optimizing the fabrication parameters to mode match the Ti-diffused and proton-exchanged waveguides, choosing the proper polarizer length, and placing the proton-exchanged section at the output of the sample to eliminate depolarization effects in the Ti-diffused waveguides.

References

1. P.G. Suchoski, T.K. Findakly, and F.J. Leonberger, "Depolarization in Ti: $LiNbO_3$ waveguides and its effect on circuit design," Submitted to Electron. Lett.

2. J.J. Veselka and G.A. Bogert, "Low-loss TM-pass polarizer fabricated by proton exchange for z-cut Ti: $LiNbO_3$ waveguides," Electron. Lett. 23, p.29-30 (1987).

3. J. Ctyroky and H.J. Henning, "Thin-film polarizer for Ti: $LiNbO_3$ at $\lambda = 1.3$ μm," Electron. Lett. 22, p.576-577 (1986).

4. T.K. Findakly and B. Chen, "Single-mode transmission selective integrated optical polarizers in $LiNbO_3$," Electron. Lett. 20, p.128-129 (1984).

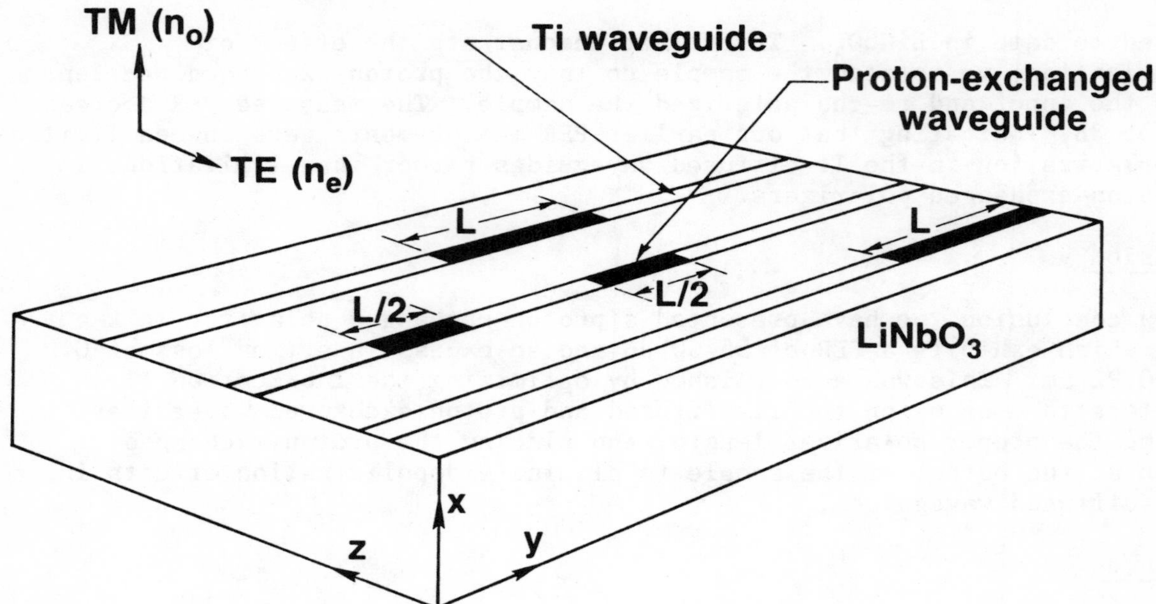

Figure 1. Three different geometries used for proton-exchanged polarizer.

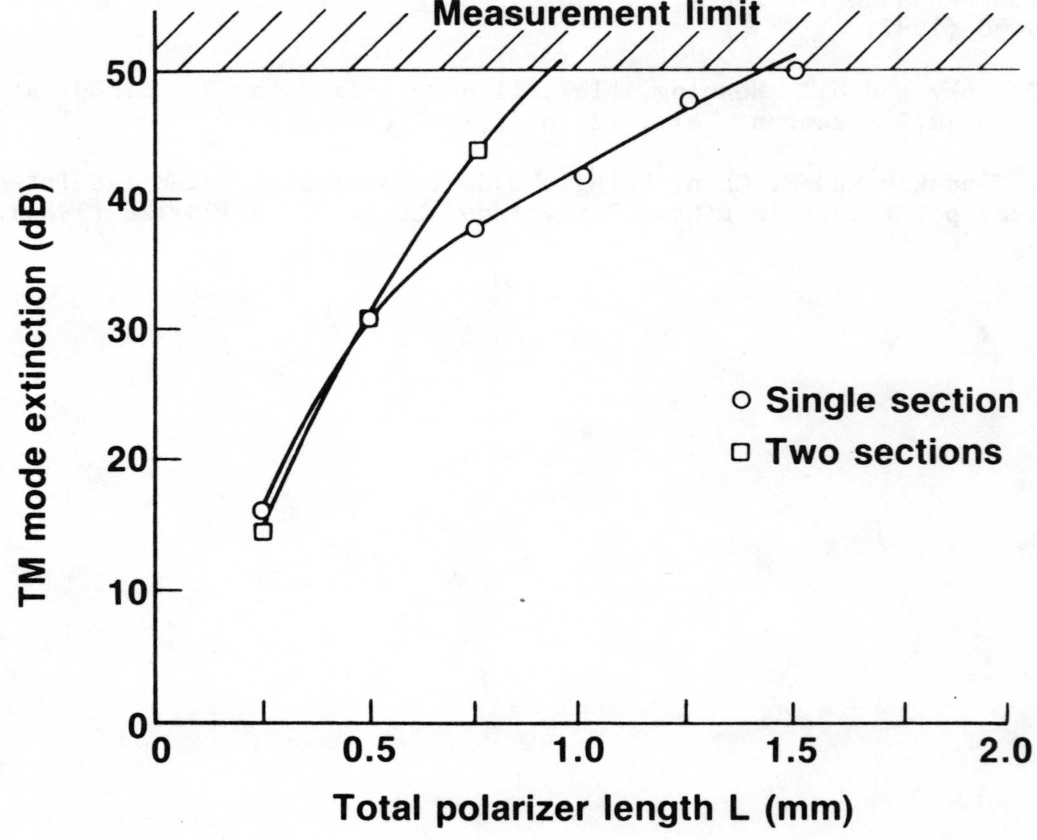

Figure 2. Measured TM mode extinction for one and two section proton-exchanged polarizers located in middle of sample.

THURSDAY, JANUARY 28, 1988

MEETING ROOM 2-4-6

3:30 PM–5:00 PM

ThEE1–5

SENSOR COMPONENTS

Mokhtar Maklad, EOtec Corporation, *Presider*

LEAD-INSENSITIVE OPTICAL FIBRE SENSORS (INVITED PAPER)

John P Dakin, Plessey Research (Roke Manor) Ltd,
Romsey, Hants, U.K., SO51 0ZN

ABSTRACT

The paper describes the methods by which optical fibre sensors, intended for remote measurement, may be rendered insensitive to physical perturbation of the sensor leads and to any changes that could otherwise occur due to remating of connectors or lead replacements.

Summary

The most important aspect of a sensing system, assuming it has achieved its primary purpose of having a stable and well-behaved response to the desired measurand, is that it should have a low sensitivity to any environmental changes which it is not required to measure. The design of the sensor head (or sensing regions of fibre, in the case of an intrinsic sensor) will, in any well designed system, ensure that undesirable changes in the sensing zone do not adversely affect the performance. However, the sensing region is frequently located away from the monitoring station and will be connected by one or more optical fibre leads. This paper discusses, firstly the main types of perturbation that can occur in practical systems, secondly what sensing methods may be used to overcome the problems, and finally which second order errors may still remain, if care is not taken with the design.

In this short summary, it has not been possible to give a full overview of all the problems that can occur and therefore a number of the more important effects and their potential solutions are shown in fig (1). In the presented paper, the effects and a number of published approaches to the problem will be reviewed in more detail and several examples will be given of practical sensor systems where lead insensitivity has been taken into account.

TABLE 1

CABLE PERTURBATIONS AND CONNECTION PROBLEMS	**EFFECT ON SENSOR SIGNALS**	**POTENTIAL SOLUTIONS**	**RESIDUAL PROBLEMS THAT MAY GIVE RISE TO SECOND ORDER ERRORS**
Changes in Attenuation (e.g. micro- and macro-bending induced) Changes may also occur due to lead replacement or insertion of additional connectors or extension leads	Amplitude Modulation (Note that this may also have a weak wavelength dependence)	Avoid simple intensity measurement methods.Use wavelength filtering(1) time delay sensors(2) two-wavelength referencing (3) etc.	Very few if well designed. Two-wavelength referenced schemes may show errors if wavelengths not close together.
		Balanced Optical Bridge and other reference path methods	(i) The reference paths may still be affected if care is not taken (ii) The beam splitters and combiners may themselves be environmentally sensitive
Mode Conversion in Cable (Strain - induced)	Minor Amplitude Modulation effects (modal noise)(5)	Use large diameter fibre and broadband sources to prevent modal noise. Avoid unreferenced amplitude sensing concepts.	Less of a problem than microbending losses, except when mode-dependent connections, couplers, or spatial filters are present.
	May cause changes in coupling ratio in balanced bridge systems.	Use Couplers which are mode-independent.	Most practical splitting components exhibit some degree of mode dependence.

Polarisation Modulation due to bending and strain-induced birefringence	A particularly serious problem in polarimetric sensors.	Polarimetric sensors may be remoted using leads of polarisation - maintaining fibre to feed in plane-polarised light and guide back the orthogonal components of the output light to be remotely analysed (6)	Polarisation mode conversion is likely to occur in the leads unless they are well isolated from strain
	Amplitude modulation could occur if polarisation-dependent components eg beamsplitters are present in the system.	Use multimode fibre for intensity based sensors as otherwise problems could occur even with two-wavelength referenced systems.	Not likely to be a significant problem in multimode systems.
Slow-Acting Optical Path Length changes due to temperature fluctations.	Modulates optical phase slowly and may bias coherent sensors from desired quadrature point.	Use a fringe counting scheme (7) or heterodyne detection (8) in a coherent sensor arrangement. Use a thermally-balanced interferometer (9)	Very few if care taken with signal processing scheme, particularly when effect to be monitored shows rapid variations with respect to the thermal effects.
Faster-Acting optical Path Length Changes Due to Vibration in leads.	May create interferring signal in hydrophone systems	Arrange probe and reference beam of interferometer to pass through connecting leads.(10) Design cable to reduce vibration effects	There may be small second order effects if the sample and reference signals do not pass simultaneously through the leads.

Connector interfaces	Connectors may form low-finesse Fabry-Perot filters, which could result in spectral variations in loss	Use wavelength-filtering methods with a single narrow band output(1)For two-wavelength -referenced schemes, use large diameter fibres.	Not likely to be a major problem with multimode fibre systems.
	Connector surfaces may become contaminated. May cause loss or even spectral filtering.	Use wavelength filtering with a single narrow band output. Avoid unreferenced amplitude methods.	Two-wavelength referenced scheme could suffer errors if contamination has different transmission at two wavelengths.

References

1. Dakin J P, "Analogue and Digital Extrinsic Optical Fibre Sensors based on Spectral Filtering Techniques", Proc. Fibre Optics '86 London SPIE 468 (1984) pp 219 - 226.

2. Johnson M, Ulrich R, "Fibre Optical Strain Gauge" Electron Letts 14 (1978) pp 432 - 434

3. Morey W, Glen W H et Al, NASA Report CR - 165125 (1981)

4. Giles I P, Uttam D, McNeil S, Culshaw B, "Self Compensating Technique For Remote Fibre Optics Intensity Modulated Sensors". Proc Fibre Optics '85 London. SPIE 522 (1985) pp 233 - 239

5. Epworth R E, "The Phenomenon of Modal Noise in Analogue and Digital Optical Fibre Systems" Proc 4th ECOC, Genoa Sept 1978 pp 492 - 501.

6. Varnham M P et Al, "Polarimetric Strain Gauges Using High Birefringence Fibre" Electron, Letts 19 pp 699 - 700.

7. Sheem S K, Giallorenzi T G, Koo K P, "Optical Techniques to Solve The Fading Problem in Fibre Interferometers". Appl Optics 21 (1982) p 689.

8. Hall T J, Phd Thesis Univ. of London U.K. (1980)

9. Dakin J P, Wade C A, "Compensated Polarimetric Sensor Using Polarisation-Maintaining Fibre in a Differential Configuration" Electron Letts 20 (1984) pp 51 - 53

10. Dakin J P, Wade C A, Withers P B, "Engineering Improvements to Multiplexed Interferometric Fibre Optic Sensors" Proc Fibre Optics '85 London SPIE 522 (1985) pp 226 - 232.

ThEE2-1

Surface Plasmon Resonances in Thin Metal Films for Optical Fibre Devices

G Stewart, W Johnstone, B Culshaw and T Hart*
Department of Electronics and Electrical Engineering
The University of Strathclyde
Glasgow G1 1XW

*Sifam Ltd, Woodland Road, Torquay TQ2 7AY

Introduction

Surface-plasmon polaritons in metal films may be used to construct optical fibre devices such as polarisers, polarisation selective couplers and also offer possibilities for improving the efficiency of certain sensor and non-linear devices due to the enhanced surface fields associated with the surface plasmon.[1,2] We report here theoretical and experimental studies of surface plasmon resonances excited in thin metal films deposited on optical fibres which have been polished to reduce their cladding thickness to a few microns (or less).

Dispersion Relations

We consider the case of an asymmetrically bounded planar metal film of thickness, t, with dielectric media on either side of indices, n_1 and n_3 (n_1, n_3 real and $n_3 \geqslant n_1$). In order to obtain some useful simplified expressions we assume, initially, an ideal (loss-less) metal of index, $n_2 = jk$. The effective index, n_e, of a mode in the structure is given by:[1,2]

$$k_o \cdot t\sqrt{n_e^2 + k^2} = \tanh^{-1}(A_1) + \tanh^{-1}(A_3) \tag{1}$$

Where $k_o = 2\pi/\lambda$, $n_e = \beta/k_o$, β is the propagation constant of the mode and A_1, A_3 are given by:

$$A_i = \frac{k^2}{n_i^2} \cdot \sqrt{\frac{n_e^2 - n_i^2}{n_e^2 + k^2}} \tag{2}$$

Equation (1) may be rewritten as:

$$bT = \tfrac{1}{2} \ln \left[\frac{(A_1 + 1)(A_3 + 1)}{(A_1 - 1)(A_3 - 1)} \right] \tag{3}$$

where $b = \sqrt{(n_e^2 + k^2)}$ and $T = k_o t$.

Since $bT > 0$, then $(A_1 + 1)(A_3 + 1)/(A_1 - 1)(A_3 - 1) > 1$. Several regions of operation may be identified from equation (3) as illustrated in the computed dispersion curves in Figure 1:

(i) For A_1 or $A_3 = 1$ ($T \rightarrow \infty$), we obtain the decoupled surface plasmons at either boundary.

(ii) For $n_e > n_3$ (T finite), we obtain two bound modes (symmetric and antisymmetric) and two leaky modes[1].

(iii) For $n_e < n_3$, we have a mode leaking into the n_3 dielectric at an angle $\theta = \sin^{-1}(n_e/n_3)$.

(iv) For $n_e = n_3$, we derive an expression for the cut-off thickness, T_c, of the bound mode ($A_3 = 0$ in equation (1)):

$$T_c = (n_3^2 + k^2)^{-\frac{1}{2}} \tanh^{-1} \frac{k^2}{n_1^2} \cdot \sqrt{\frac{n_3^2 - n_1^2}{n_3^2 + k^2}} \qquad (4)$$

Typically, for a symmetrically bounded metal film of index $n_2 = j12$, equation (1) yields a bound mode with effective index $n_e = 1.454$ for $T = 0.12$ ($t \simeq 240Å$ at $\lambda = 1.3\mu m$, $n_1 = n_3 = 1.45$). For an asymmetrically bounded film, the mode has a cut-off thickness $T_c = \frac{1}{2}T = 0.06$ ($t \simeq 120Å$, $n_3 = n_e = 1.454$, $n_1 = 1.45$).

Experimental Coupling to Surface Plasmons

Thin metal films of silver or aluminium ($n_2 \simeq 0.3 + j9$ and $1.2 + j12$, respectively, at $\lambda = 1.3\mu m$) were deposited on the polished region of optical fibres and coupling to surface plasmons was observed by measurement of the TM/TE polarisation extinction ratios in the fibre output at $\lambda = 1.3\mu m$. Figure 2 shows typical results obtained for an Al film, where extinction ratio is plotted against the superstrate (liquid) index, n_3, for various film thicknesses (Å). Two distinct regions are apparent, as predicted by the dispersion relations:

(i) For $n_3 < n_{ef}$ and $100Å < t < 300Å$ coupling to a resonance occurs with an optimum thickness at $t \sim 250Å$.

(ii) For $n_3 > n_{ef}$ and $t < 130Å$, coupling to a leaky wave occurs, with energy leaking into n_3 dielectric.

(n_{ef} is the effective index of the guided mode in the fibre).

We have used the above principles to construct efficient fibre polarisers (> 50dB extinction ratio) and polarisation selective couplers. Application is also being made in the design of sensor and non-linear devices.

Conclusion

We have derived some simplified expressions describing bound and leaky surface plasmons in thin metal films. These results have been used to construct efficient optical fibre polariser devices and have important applications in the design of sensor and non-linear devices exploiting surface plasmon phenomena.

References

1. G I Stegeman et al, Opt Lett, 8, p383, 1983.

2. G Stewart et al, "Design of optical fibre polarisers", Proc SPIE, 734, 1987.

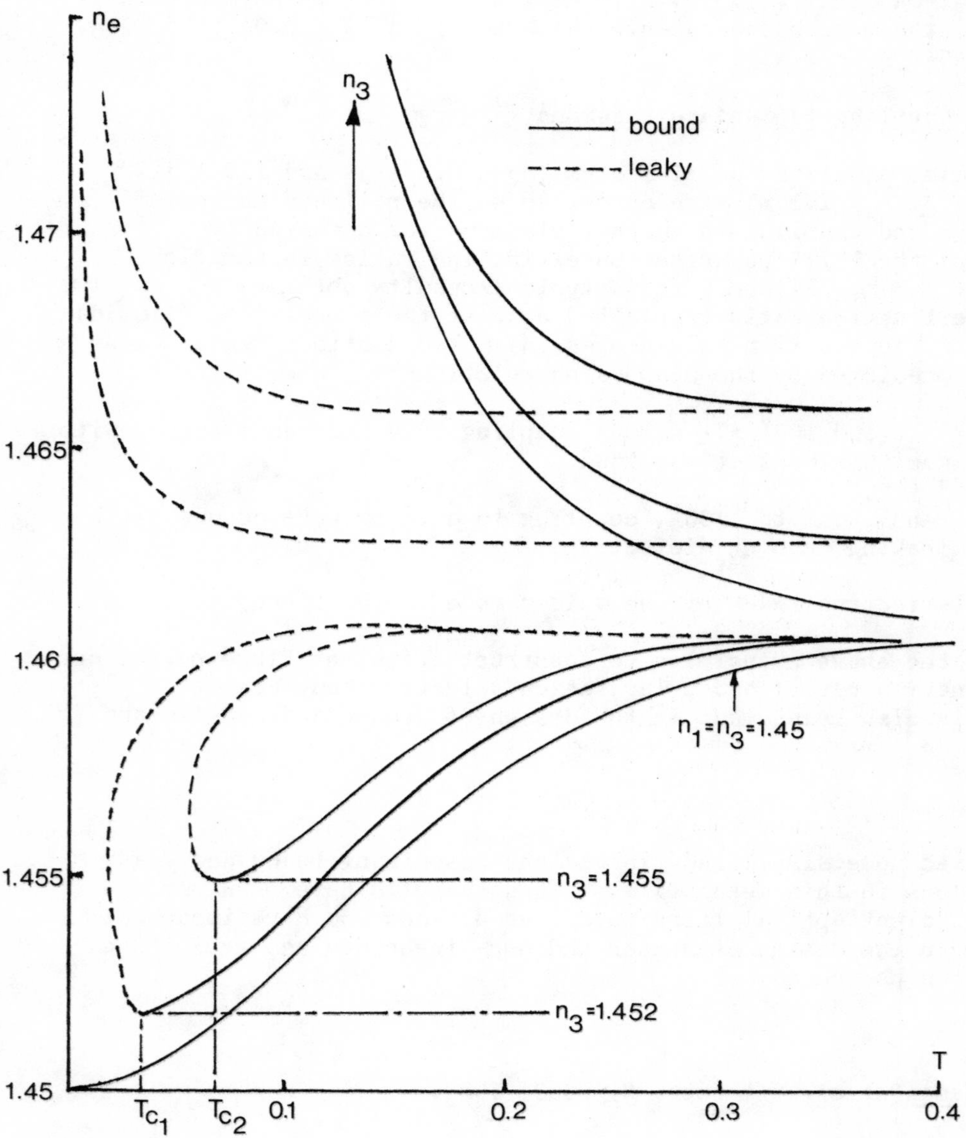

FIG.1 THEORETICAL DISPERSION CURVES (n_1=1.45, n_2=j12)

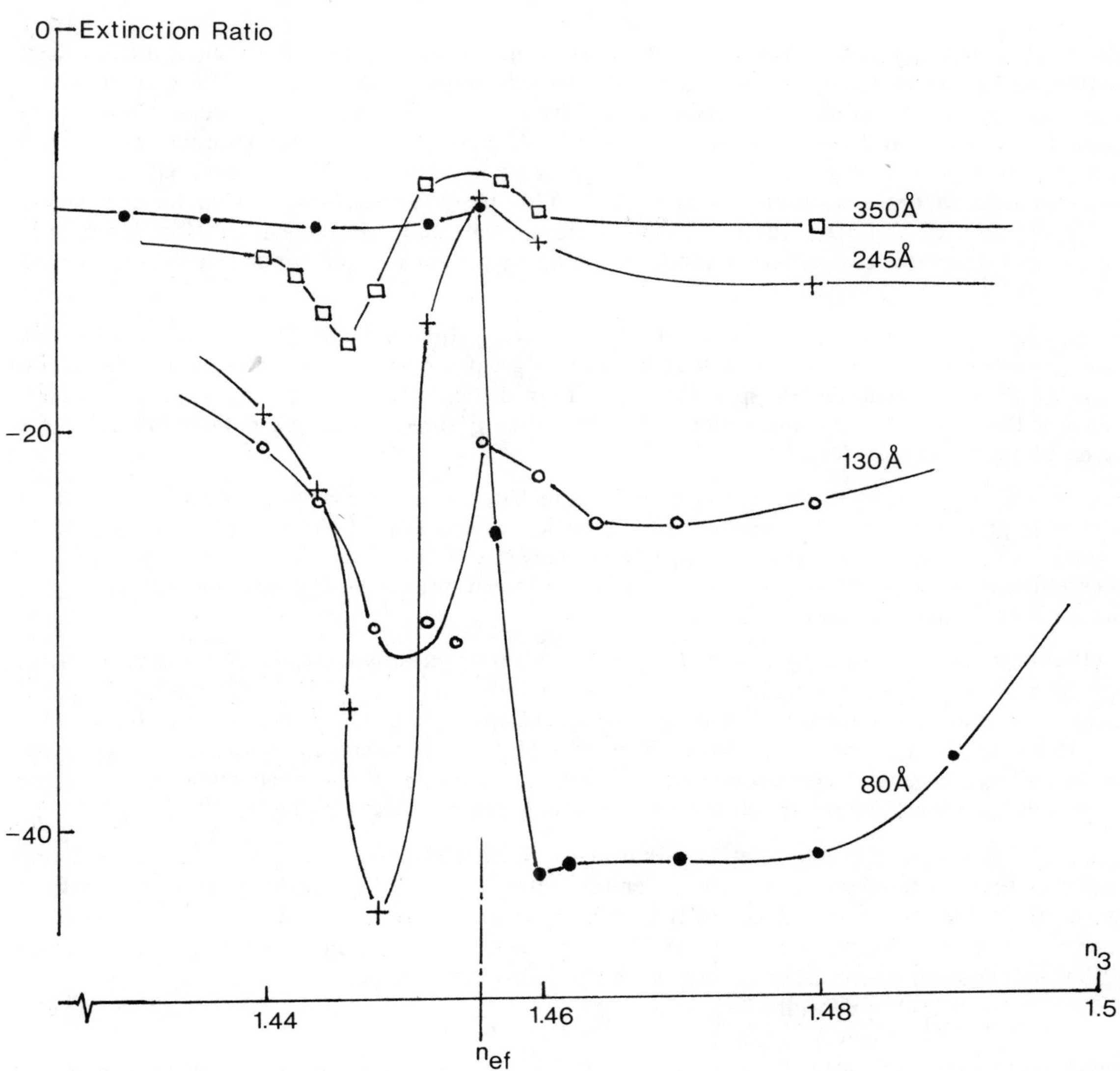

FIG.2 EXPERIMENTAL COUPLING TO PLASMONS

WAVELENGTH CONVERSION FOR ENHANCED FIBER OPTIC ULTRAVIOLET SENSING

R. A. Lieberman
L. G. Cohen
V. J. Fratello
E. Rabinovich

AT&T Bell Laboratories
Murray Hill, New Jersey 07974

A fluorescent coating (phosphor) applied to the tip of an optical fiber converts incident radiation to longer wavelengths, and can considerably improve fiber-based UV spectrometry. Longer wavelengths are more readily transmitted by the fiber, are less apt to cause UV-induced damage in the fiber, and are more easily detected. Alignment of the fiber is much less critical for fluorescent-tipped fibers and the use of inorganic phosphors which have "peaked" excitation curves can help increase wavelength selectivity. This paper presents a method for evaluating the effectiveness of different combinations of fiber, phosphor, source collimation state, and detector, and describes two different methods for applying durable phosphor coatings to optical fibers.

High transmission loss, caused by Rayleigh scattering (proportional to $(1/\lambda)^4$) and (for wavelengths below 185 *nm*) the Urbach absorption edge, presents the most formidable barrier to the use of fibers in remote UV spectroscopy. This difficulty, illustrated by Figure 1 (which presents a loss curve for a commercial "UV transmitting" fiber), was the main motivation for work on phosphor-tipped fibers.

The total efficiency, η_p, for conversion of optical energy from one wavelength to another by a phosphor is proportional to the product of the optical absorption of the phosphor (measured at the input wavelength) and the quantum efficiency for conversion of absorbed photons to fluorescent emission photons. The relationship between input energy and output energy is linear for most practical cases.[1] [2]

The efficiency for coupling source light to guided modes in phosphor-tipped fibers is $\eta_T = \eta_p \eta_c$, where $\eta_c = (NA / n_m)^2$, the coupling efficiency for a Lambertian source, [3] independent of fiber alignment or source collimation (NA is the numerical aperture of the fiber, and n_m is the index of refraction of the phosphor medium on the fiber tip). The maximum coupling efficiency for bare fibers, η_B, is near 1 for collimated sources, and $\eta_B = (NA)^2$ for properly aligned diffuse sources, but η_B depends strongly on the incidence angle of the incoming light.

A quantitative measure of the relative advantages of various phosphor-tipped fibers, including all of the above considerations, is the "enhancement length," l_e, defined as the length of detector fiber for which the fluorescent-tipped approach becomes a more efficient means of transmitting optical energy than the use of a bare fiber. The enhancement length depends on the fiber loss characteristic, $\alpha(\lambda)$, as shown by the following relation:

$$l_e(\lambda_1, \lambda_2) = \frac{10\log(\eta_T / \eta_B)}{\alpha(\lambda_2) - \alpha(\lambda_1)}$$

where λ_1 is the the incident (UV) wavelength, and λ_2 is the fluorescent emission wavelength of the phosphor coating. An integral average of the above expression can be used to define l_e for broadband sources and phosphors.

Table I lists enhancement lengths achievable by applying typical phosphors[1] [2] to the tip of a high NA fiber with a loss curve similar to that shown in Figure 1. Lower values of l_e indicate situations for which the phosphor-tip approach is more advantageous. Systems which are used for detection of diffuse UV sources are the most aided by the phosphor-tipped fiber approach.

TABLE I: ENHANCEMENT LENGTH (l_e) In Meters										
	DIFFUSE SOURCE Wavelength (nm)					COLLIMATED SOURCE Wavelength (nm)				
PHOSPHOR	150	190	200	225	254	150	190	200	225	254
NBS No. 1026	-	4.6	5.2	10.6	13.9	-	6.8	7.4	15.6	21.7
NBS No. 1027	-	4.5	5.2	10.9	13.6	-	6.4	7.4	15.6	21.7
NBS No. 1028	-	4.9	4.8	10.6	15.5	-	6.8	6.8	15.6	23.3
Liumogen	0.0	6.0	6.6	13.4	18.6	0.0	8.2	8.9	17.8	26.7
Coronene	0.0	5.7	6.6	13.4	18.0	0.0	7.7	8.9	17.8	25.0

For practical UV monitoring systems, the response of the detector is often the determining factor in system performance. To account for detector differences in phosphor-tipped fiber comparisons, a "total enhancement length", L_e, can be defined:

$$L_e = l_e + \frac{10\log(D_2(\lambda_2)/D_1(\lambda_1))}{\alpha(\lambda_2) - \alpha(\lambda_1)}$$

where $D_1(\lambda)$ is the response function for the detector used in the bare-fiber system, and $D_2(\lambda)$ is the response function for the detector used in the phosphor-tipped fiber system. In most practical situations, $L_e < l_e$.

Table II shows total enhancement lengths for diffuse sources in systems incorporating two different "UV-optimized" detectors.

TABLE II: ENHANCEMENT LENGTH (L_e) Including Detector Response DIFFUSE SOURCE										
	PHOTOMULTIPLIER DETECTION Detectors: Hammamatsu R1259(UV) Hammamatsu R928 (VIS) Wavelength (nm)					SILICON P-I-N DETECTION Detector: PAR HUV 2000B Wavelength (nm)				
PHOSPHOR	150	190	200	225	254	150	190	200	225	254
NBS No. 1026	-	4.2	4.8	9.4	8.6	-	1.4	3.1	7.1	9.7
NBS No. 1027	-	4.3	5.0	10.0	8.8	-	1.1	2.9	6.9	8.5
NBS No. 1028	-	4.7	4.6	9.8	10.9	-	1.1	2.2	6.1	9.7
Liumogen	0.0	5.8	6.4	12.7	14.1	0.0	2.2	3.8	8.5	12.1
Coronene	0.0	5.5	6.4	12.5	13.2	0.0	2.3	4.3	9.4	12.9

A wide variety of high-efficiency phosphors are available commercially. We affixed two different powdered inorganic phosphors to optical fiber tips, using a variation of a method used to prepare cathode-ray tube faceplates. The phosphor was bound to the end of the fiber with a "spin-on glass", a thermosetting polysiloxane polymer resin that can be pyrolyzed to form an inorganic glass. The resin was suspended in an organic solvent and a droplet of the resulting slurry was applied to the tip of the fiber. After the slurry dried, the fiber tip was baked to hardness to remove any remaining organics and water and to fully vitrify the spin-on glass.

Excitation and emission spectra from fibers prepared by the above method were obtained with a scanned monochromator system connected to photon counting electronics. The spectra obtained from two different phosphors applied in this fashion are shown in Figures 2(a) and 2(b). These spectra clearly show that fluorescence from phosphors applied to fiber tips can be directly coupled to light-guiding modes in the fiber core. The sharp emission spectrum of the phosphor response in Figure 2(a) illustrates the possibility of using a phosphor to convert broadband UV light to narrow-band visible light for possible multiplexing applications. The double excitation peak of the phosphor in Figure 2(b) illustrates the possibility of "rejecting" the sharp 254 *nm* mercury emission peak while collecting ultraviolet radiation in other regions of the 200 - 300 *nm* band.

Another inorganic phosphor which can be applied to form a hard coating on quartz is terbium-doped yttrium orthosilicate (Tb/YOS). Figure 2(c) shows the spectrum of guided fluorescence emission from a 1mm diameter quartz rod coated with Tb/YOS using a sol-gel method,[4] and illuminated from the side with 254 *nm* light.

Fusing fluorescent species to optical fibers presents a viable method for collecting ultraviolet light for transmission to a remote detector. The ability to use a side-coated, side-illuminated geometry as well as a coated-tip, end-illuminated geometry demonstrates a flexibility which may prove useful in wide-area monitoring applications where it is desired to collect emission from diffuse sources. The techniques presented in this paper allow a great deal of latitude in the choice of lightguide geometry, phosphor, and processing methodology, and are useful in creating optical waveguide probes for remote UV sensing, particularly at vacuum ultraviolet (VUV) and extreme uiltraviolet (XUV) wavelengths.

REFERENCES

1. Bril, A., **Luminescence of Organic and Inorganic Materials**, Kallmann, H.P. and Marmor-Spruch, G. (eds), Wiley, N.Y., p. 479, (1967).

2. Viehman, W., Butner, C.L., Cowens, M.W., *SPIE Proc. 289*, p. 146, (1981).

3. Snyder, A. W., Love, J. D., **Optical Waveguide Theory**, London, Chapman and Hall, 1983.

4. Rabinovich, E. M.; Schmulovich, J.; Fratello, V.J.; *J. Am. Ceram. Soc*, submitted May, 1987.

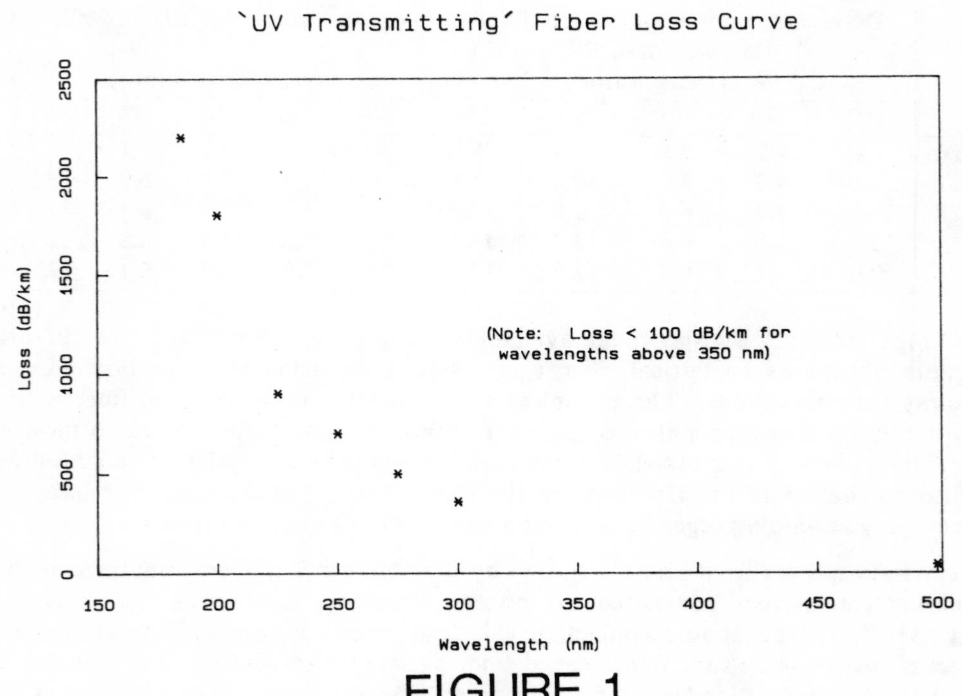

FIGURE 1

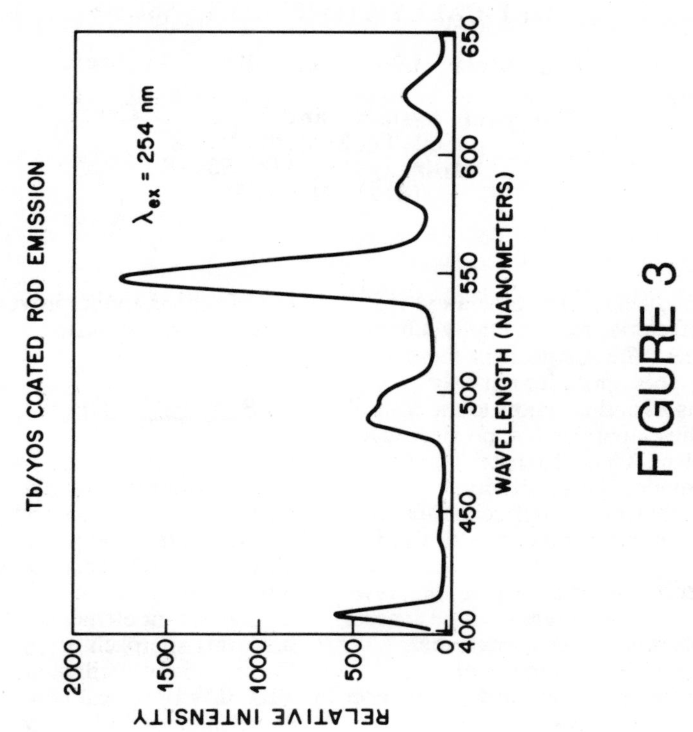

FIGURE 3

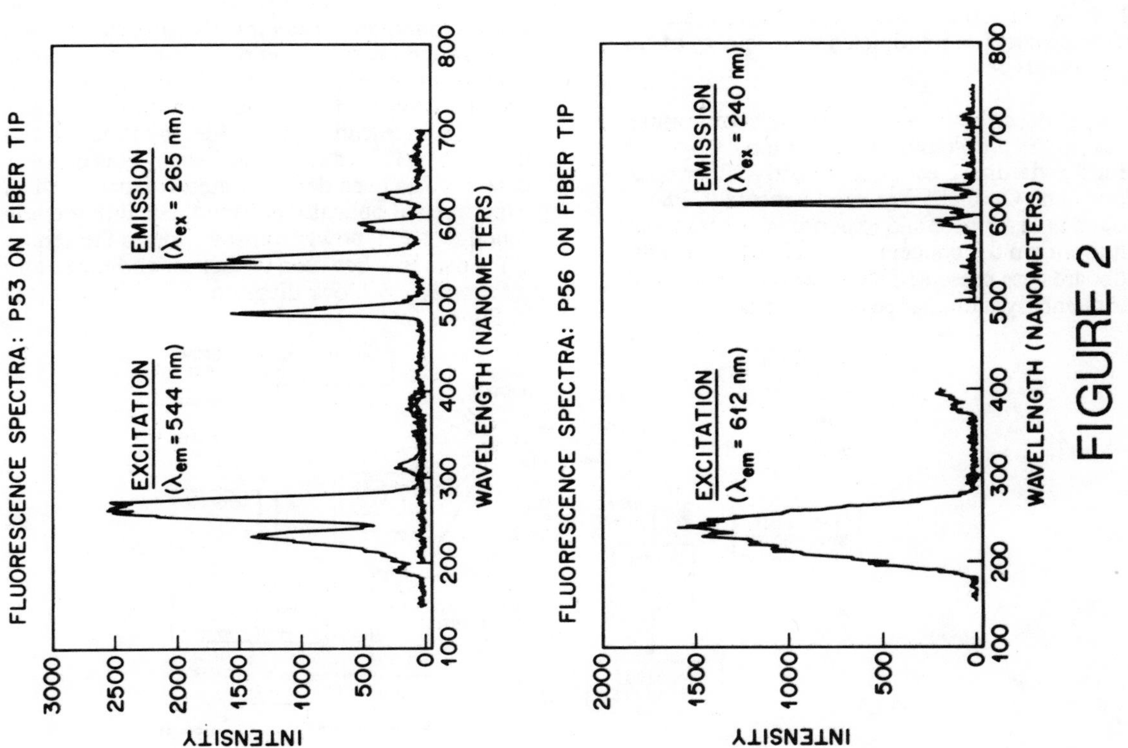

FIGURE 2

ThEE4-1

OPTICALLY POWERED SENSORS

Paul Bjork, James Lenz, Kyuri Fujiwara

Honeywell Systems and Research Center
3660 Technology Drive
Minneapolis, MN 55418
(612) 782-7638

INTRODUCTION

Many advanced systems such as next generation aircraft and automated factories require highly reliable, precise control functions. The designers of these systems are specifying fiber optics for not only interconnecting the sensors and actuators to the control circuits but also providing a passive (i.e. no electrical power) sensor mechanism. Optical sensors that have been developed can provide high sensitivity, EMI/EMP immunity, light weight, high reliability, low cost, and a single fiber for input and output of signals.

Although fiber optic sensors offer great potential, few products are yet available. The failure of these sensors to meet some other basic requirements limits their success in the marketplace. These requirements include the absolute measurement of a single parameter with very low sensitivity to other environmental parameters, and wide dynamic range and high linearity.

Perhaps one of the most important requirements in building a system of sensors is the incorporation of a single standardized interface which uses the same interface protocol to multiplex a wide variety of fiber optic sensor types.

This paper describes a very near-term compromise solution to the problem of design of totally optical, system-standardized, accurate, sensitive fiber optic sensors. The Optically Powered Sensors (OPS) approach is described, and experimental breadboard results confirm the concept. Several different sensor breadboards are presented along with experimental linearity and dynamic range data. Further modifications and improvements to the system are presented in conclusion.

OPS APPROACH

In the optically powered sensors approach, sufficient optical power is transmitted to the remote sensor via the fiber link so that a usable level of optically generated electrical power is available for data acquisition and transmission. The wide variety of reliable, inexpensive electronic transducers can then be used as sensor elements. Several recent developments drive this approach beyond the "nice idea" stage. Commercially available laser diodes are able to supply >100 mW of optical power into the fiber at a wavelength of 850 nm. A new generation of CMOS micro-power analog to digital (A/D) converters, developed at Honeywell, consume only microwatts of power. These are key components in a system in which a single 100 μ core multimode optical fiber is used to provide optical power, to address one of several multiplexed, electrically isolated EMI shielded sensors and modules and provide a data return path. A set of matched photodetectors arranged in series are used to convert the incoming light to a DC voltage. Data communication is provided by standardized "off-the-shelf" LED's and photodiodes. This experimental data link has been demonstrated at Honeywell. Our experimental optically powered data link provides two functions: (1) power transmission to the sensor, and (2) a data link between the sensor and the control electronics. A block diagram of this system is shown in Figure 1.

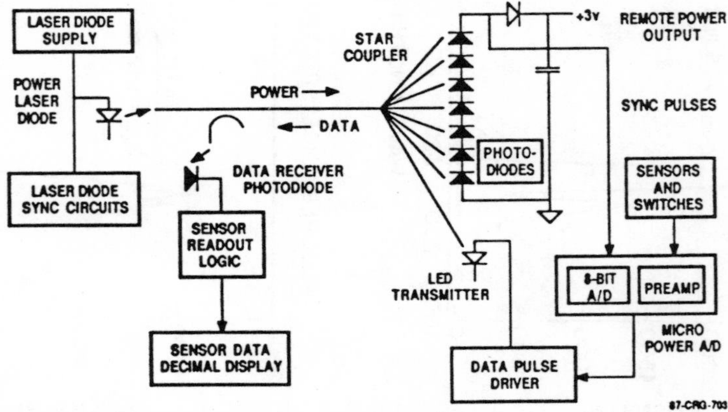

Figure 1. Optically Powered Sensors Data Link Demonstrator Schematic

A description of the schematic is as follows. A high-power Spectra Diode Labs SDL-2420 laser diode provides both DC power and synchronization signals across the single 100 μ fiber. The fiber link consists of 130 meters of fiber and three AMP 501068 fiber connectors, with typical loss less than 2 dB per connection. An Amphenol 1 x 8 coupler provides power splitting to seven Honeywell SD3478 photodiodes and one HFE 4000 LED at 820 nm. Current from the photodiodes is stored on a 220 uF, 5 volt electrolytic capacitor, which supplies about 3.5 volts for sensor electronics. Sensor data is transmitted serially by the LED, and recovered using an Canstar 2 x 2 coupler and a Honeywell Schmitt Trigger photodiode capable of resolving the 1 μS, 2 μW pulses.

The key to single-fiber operation is the standardized synchronization protocol for power supply and communication between the interface module and the remote sensor. The communication process is initiated by a polling operation at the interface module. Logic in each sensor on the network is wired to respond to a unique signal or a address on the incoming optical power. This signal may be a particular combination of pulses or a particular frequency or repetition rate of pulses. All the advantages of low power CMOS digital circuitry are available for address coding and decoding.

When the sensor is triggered in this way, the sensor A/D converter is activated and the 8 bit laser result is sent back to the interface module one bit at a time between a set of 25 short optical diode pulses, which now are interpreted as synchronization pulses. Since the sensor transmits the only optical power on the network between synchronization pulses, signal to noise ratio at the interface module is maximized. Figure 3 is a timing diagram of the process. Information from the last sensor addressed is sent bit by bit between synchronization clock pulses. Laser retroreflections into the detector are easily masked out using this synchronization scheme.

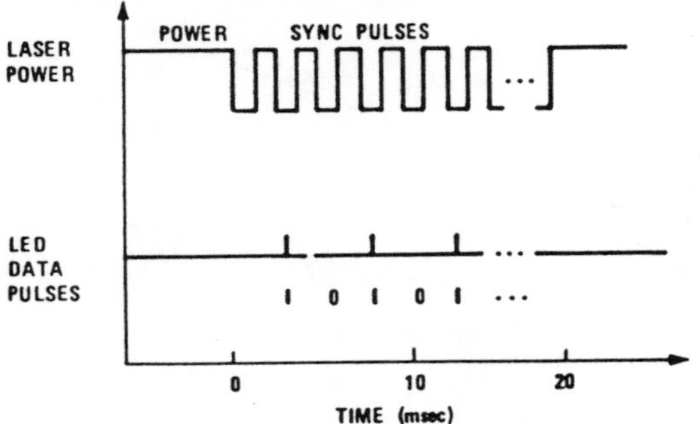

Figure 2. Protocol Timing Diagram Example.

The overall efficiency of converting the optical power generated by the laser diode into electrical power at the sensor is approximately 1.8%. Table 1 lists the power budget for the energy conversion path outgoing to the sensor, as well as the optimum efficiency that could be achieved. For the present experimental configuration, 1-2 mW of electrical power is available at the sensor.

TABLE 1. OUTGOING POWER BUDGET

COMPONENETS	OUTPUT POWER (mW)	EFFICIENCY (%)
1. LASER DIODE	100.00	10.00
2. SPLICE	95.50	95.50
3. READOUT COUPLER	38.02	39.81
4. SPLICE	36.31	95.50
5. MULTIMODE FIBER	22.91	63.10
6. SPLICE	21.88	95.50
7. STAR COUPLER	13.80	63.10
8. SPLICE	13.18	95.50
9. OPT/ELEC CONVERSION	1.78	13.49
10. OVERALL EFFICIENCY		1.8

Table 2 then lists the power budget returning from the sensor LED back to the power/data interface module. Powers listed in this table are peak powers. For the experimental prototype, a sensor reading of 8 bits of data is transmitted once each second. Thus, transmission consumes less than 5 microwatts of average power. This results in 3-4 μW peak power at the receiving photodiode in the data receiving module. Thus, it is possible to supply sufficient power to the remote module as well as to actually communicate the information.

TABLE 2 RETURNING POWER BUDGET

COMPONENT	OUTPUT POWER (microW)	EFFICIENCY (%)
1. LED	124.76	8.32E-02
2. SPLICE	119.15	95.50
3. STAR COUPLER	10.62	8.91
4. SPLICE	10.14	95.50
5. MULTIMODE FIBER	9.04	89.13
6. SPLICE	8.63	95.50
7. READOUT COUPLER	3.44	39.81
8. SPLICE	3.28	95.50
9. OVERALL EFFICIENCY		2.19E-03

Sensor electronics power utilization depends critically on the sample rate. Figure 4 shows the effect of system bandwidth, in bytes of data per second, on power dissipation requirements. It is seen that the LED power dissipation limits the present system to about 50 samples transmitted per second. Also compared is the total measured power requirement of the other elements of the magnetic proximity sensor described below.

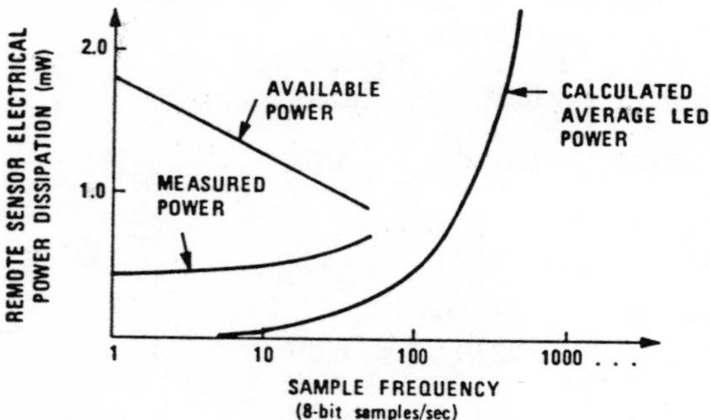

Figure 3. Power Dissipation -vs- Sample Rate

Optical-Electrical power conversion efficiency is a critical design factor. Photodiode array-current-voltage curves with initial fiber launched laser optical power as a parameter are presented in Figure 5. Plotted on the same curve is shown a load line representing the total electronic power dissipation of 327 μW for the temperature sensor described below. It is clear that the 3.0 volts required by the CMOS electronics can be supplied most efficiently by a average laser optical power of approximately 25 mW. Thus the 100 mW maximum power output of the laser diode could be divided, between as many as 4 different sensors.

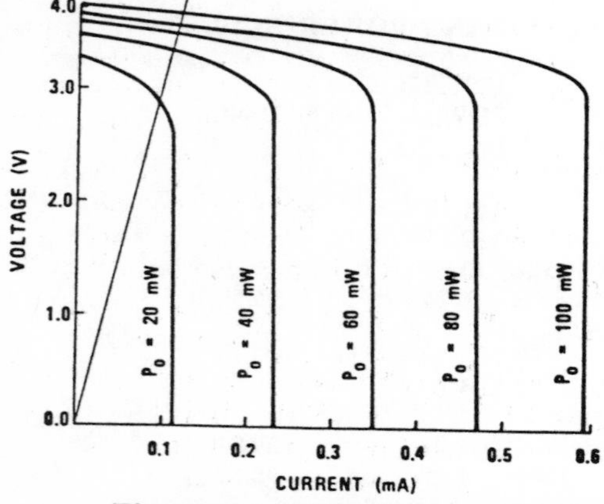

Figure 4. Photodetector Array Characteristics

APPLICATIONS

In this section we present experimental results from several sensors built using the OPS approach.

First, a magnetic permalloy resistance bridge developed at Honeywell is used in a magnetic sensor configuration. Figure 6 shows the $1/R^2$ dependence of the sensed magnetic field due to a small permanent magnet at a distance R. A distance of 1.5 cm corresponds to a 50 Gauss field. The vertical axis is the decimal display of the hexadecimal sensor readout. Total sensor end power dissipation at a sample rate of 1 Hz is 400 μW.

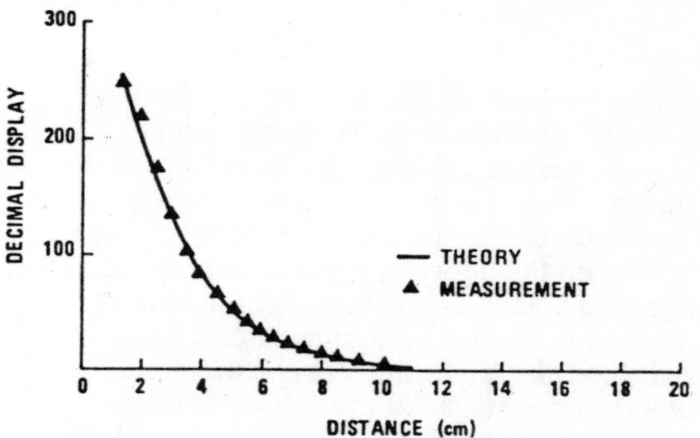

Figure 5. Magnetic Position Sensor Output

A temperature sensor is constructed with a Iron-Constantan (type J) thermocouple and an instrumentation amplifier connected to the remote electronic sensor module. Plotted in Figure 7 is a decimal display of the experimental 8-bit sensor readout. The theoretical response is also plotted for comparison. Linearity across the 100°C range can be enhanced by simple digital processing at the data receiver.

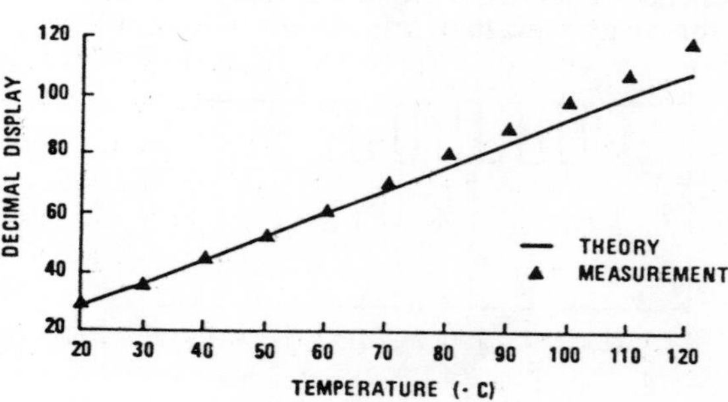

Figure 6. Temperature Sensor Output

The A/D converter linearity, as transmitted across the standardized OPS data link, is presented in Figure 8. The voltage across a variable resistor provides a simple position transducer. Plotted is resistance vs. hexadecimal display count of the data receiving interface. Linearity is seen to be within 1 %.

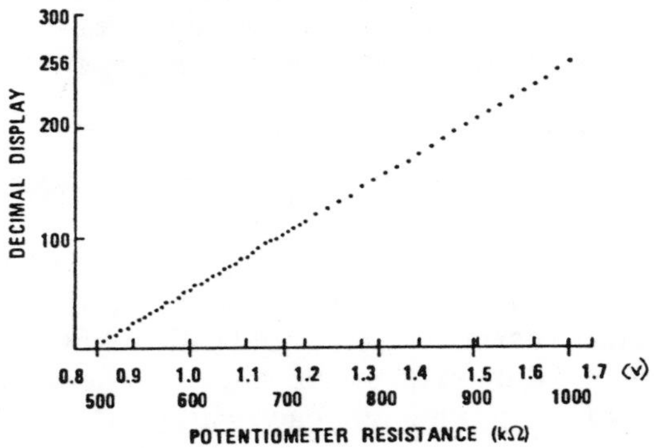

Figure 7. Sensor Electronics Linearity

High average power level tranducers have also been demonstrated using the OPS approach by using a reduced transducer duty cycle. The Microswitch SS21 magnetic switch ordinarily consumes 10-50 mW of power. The fast ~5 μs response time allows a total average sensor power dissipation of <250 μW at 25 samples per second.

DISCUSSION

Previous reports in the literature have demonstrated the operation and capabilities of optically powered sensor networks. For example, a three fiber approach has been described with one power transmitting fiber, an address fiber, and a data return fiber. We have demonstrated the ability to provide power and data communication to a variety of off-the-shelf electronic transducers using a single fiber with proper attention to synchronization issues.

In future study, we propose to breadboard a system designed to minimize optical power losses for both power transmission and the critical data return. This system will utilize a dual wavelength transmit/receive scheme at 850 and 1300 nm laser source and data wavelength, respectively. A single photodiode and pulse step-up transformer for power conversion will simplify the optics of the system. The use of a new Honeywell Schmitt detector with 10 Mbps response at only 2 μW of power further decreases the average LED power requirements. It is estimated that efficiency can be improved by more than an order of magnitude using this system, further enhancing possible system bandwidth and multiplexing capability, and enabling voice communication.

CONCLUSION

To conclude, the ability to optically power sensors is possible. The key ideas that make this feasible are the Honeywell micropower A/D converter, the development of high power laser diodes, and the state of standardization of fiber optic parts. The important benefits of this approach are:

- Modular standardization of components in a standardized optical interface.
- Ruggedization and miniaturization due to optical integration.
- The simplicity of a single fiber optic line for power transmission, multiplexing addressing, and data transmission.
- Sensor multiplexing.
- EMI protection.

These benefits make OPS a viable candidate as a systems approach to fiber optic sensing.

REFERENCES

[1] B.E. Jones, Journal Physics E Scientific Instruments September 1985, Vol. 18, No. 9, Pages 717-800.

[2] P. Hall "An Optically Powered Sensor Network", IEE Colloquium on 'Distributed Optical Fiber Sensors', Digest No. 74, May 12, 1986.

A GARNET BASED OPTICAL ISOLATOR FOR 0.8 um WAVELENGTH

G.L. Nelson, J.M. Sittig, C.H. Herbrandson, Unisys, P.O. Box 64525, St. Paul, MN 55164-0525

INTRODUCTION

Highly bismuth substituted garnet films with low optical absorption are required to achieve acceptable performance levels of <3dB insertion loss and >30 dB isolation in the 0.8 um wavelength region. A prototype optical isolator device was fabricated which has 2.7 dB insertion loss and 35dB isolation at 0.786 μm. We have prepared bismuth substituted rare-earth iron garnet films (BRIG) with 0.8 Bi/formula in a configuration in which garnet antireflection (AR) layers were included to minimize loss and backreflections from the isolator. We have also prepared single layer, single sided films with 1.3, 1.9, 2.1 and 2.3 Bi/formula.

GARNET MATERIALS

Discussion of garnet material fabrication will be covered in the order of lowest to highest bismuth content.

Multilayer garnet films were grown on (111) oriented GGG substrates (12.380A) with a structure consisting of a 1/4 wavelength thick garnet AR layer and a BRIG layer on each side of the substrate [1]. This structure permits growing BRIG films with half the thickness compared to structures with a film on a single side of the substrate but still provides a rotator which doesn't have appreciable backscatter.

The 1/4 wavelength thick AR layer is $Y_3Fe_{5-x}Ga_xO_{12}$ grown using a lithium yttrium molybdate flux [2] (LRM) to achieve slow growth rates and thus accurate film thicknesses. Growth rates and film thicknesses were measured in initial samples by AUGER depth profiles to locate the film/substrate interface and then using a profilometer to measure the depth of the crater. In this way, both films on a substrate were measured and growth conditions which gave equal and repeatable film thicknesses were determined. Ultimately, AR films with less then 10% variation between sides and from sample to sample were obtained. After establishing repeatable side to side thickness control and a series of reference samples, optical density measurements were used to measure AR film thicknesses.

BRIG films of $Bi_{0.8}Yb_{1.1}Lu_{1.1}Fe_5O_{12}$ were grown on the garnet AR layers using a $PbO-Bi_2O_3$ flux. The Faraday rotation of these films is typically 4700 ^{o}/cm at the 0.786 μm wavelength of interest (Figure 1). To achieve 45° of rotation, films of 50 um thickness per side were grown and then fine tuned by lapping. Total optical attenuation for the structure (subtracting air-surface reflections) was typically 4.3-4.5 dB. The effectiveness of the garnet AR layers is illustrated in Figure 2 which shows interference effects for two samples with relatively thin BRIG layers. The sample without a garnet AR layer has interference amplitude roughly an order of magnitude or more larger then the sample with a garnet AR layer. Based on the index of refraction for the BRIG film and the GGG substrate, we estimate that intensity reflections at the film/substrate interface are approximately -20dB without the garnet AR layer and that a sample with the AR layers has approximately -30dB intensity reflections from the two substrate surfaces.

Films of $Bi_{1.3}Gd_{1.0}Lu_{0.7}Fe_5O_{12}$ were grown on one side of Ca, Mg, Zr-substituted GGG (12.497A) using a $PbO-Bi_2O_3-B_2O_3$ flux. The Faraday rotation spectrum for these films is also shown in Figure 1. The films have typically 9100-9300 ^{o}/cm rotation at 0.786 μm. Samples were grown with approximately 50 um thickness and were lapped to achieve 45° of rotation. The Gd substitution was adjusted to reduce the magnetic moment of the samples to match the Sm-Co ring magnets used in our test devices (Figure 3). CaO and SiO_2 melt additions were made to minimize optical absorption due to Fe^{2+} and Fe^{4+} [3]. Absorption of 460 dB/cm was obtained at 0.786 and air surfaces of the samples were AR coated with

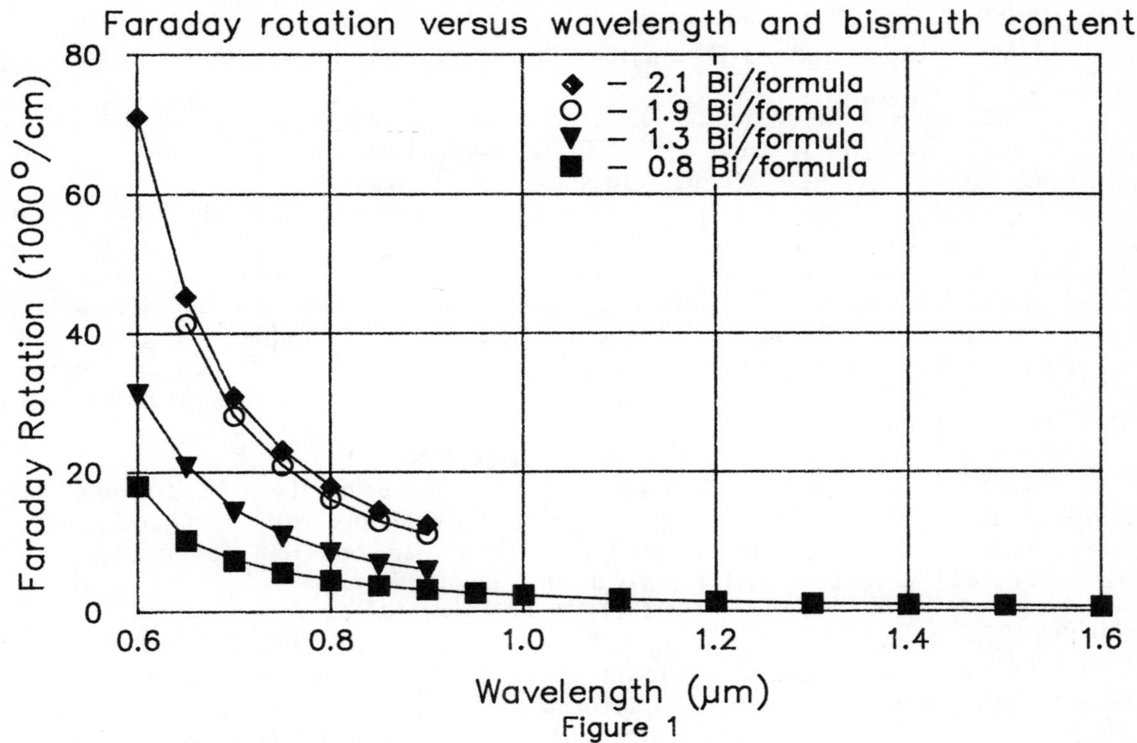

Figure 1

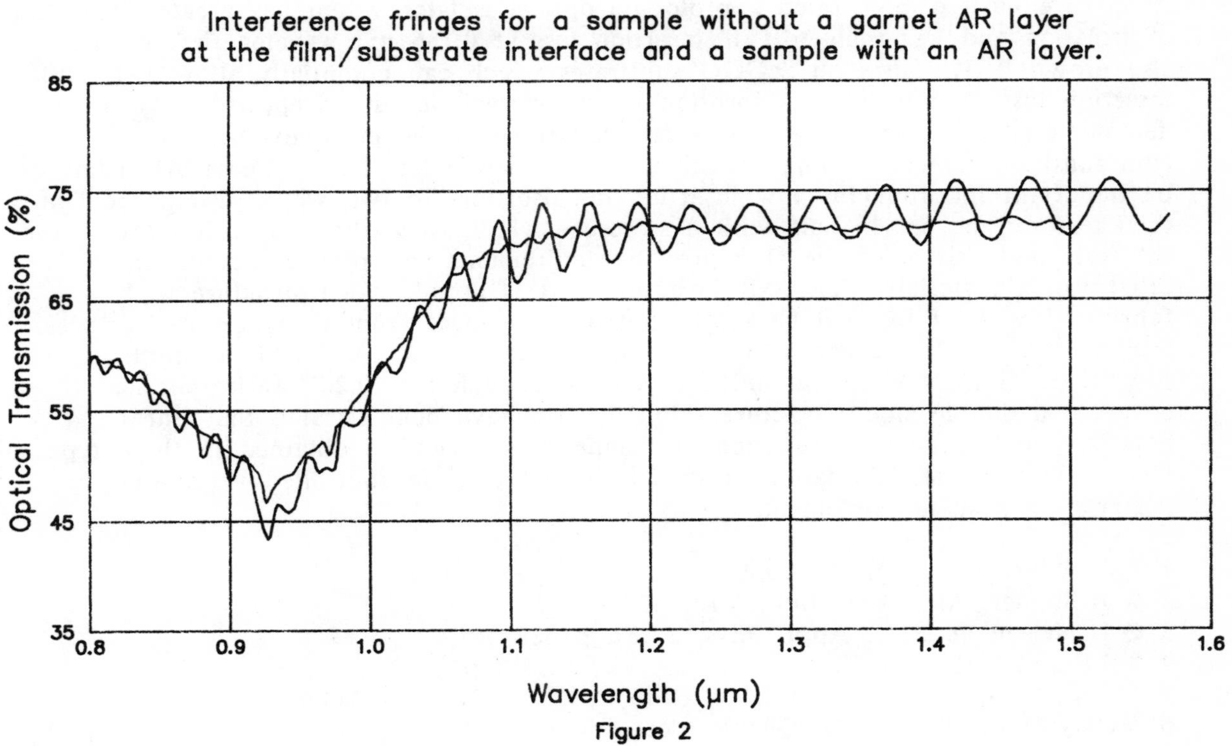

Figure 2

SiO_2 on the substrate and Al_2O_3 on the film. Insertion loss for the complete structure is 2.7dB.

BRIG films with up to 2.3 Bi/formula were grown on Sc-substituted GGG (ScGGG, a=12.566A) [4,5] using a $PbO-Bi_2O_3-B_2O_3$ flux. Faraday rotation spectrums for two of these samples are shown in Figure 1.

One sample is $Bi_{1.9}Gd_{1.1}Fe_5O_{12}$ which was grown in a single sided structure and provides 45° of rotation at 0.786 with a film thickness of 27 μm. Charge compensation to reduce optical absorption was not utilized in these films and the lowest absorption obtained is 950dB/cm. As a result the best insertion loss obtained was 2.6dB plus reflection losses. The other ScGGG based sample in Figure 1 is $Bi_{2.1}Dy_{0.9}Fe_5O_{12}$. This composition requires a film thickness of only 24 μm to provide 45° of rotation at 0.786. Thus far, however, we have only been able to grow films <10 μm thick.

ISOLATOR DESIGN

We have fabricated a prototype optical isolator using the 1.3Bi/formula material described above. The optical design is typical for isolators, with the garnet film sandwiched between polarizing prisms [6,7]. 5mm diameter Glan-Taylor prisms oriented at 45° were used and a Sm-Co ring magnet is used to bias the garnet.

ISOLATOR PERFORMANCE

Figure 4 shows the forwards and backwards insertion loss for one of the garnet films. The data was taken using 15mm diameter Glan Thompson prisms prior to device assembly as a check on the performance of the garnet at the final lapped thickness, at the 0.786 μm wavelength, and with the Sm-Co bias magnet. The performance obtained at 45° is 35.5dB of isolation and 2.75dB of forward insertion loss compared to peak values of 36dB and 2.74dB respectively for this sample.

CONCLUSIONS

We have demonstrated a prototype optical isolator which has greater then 35dB of isolation and less then 3dB of insertion loss at 0.786 μm wavelength. We have also grown BRIG films on ScGGG substrates which can potentially allow 1-1.25dB insertion loss devices when absorption reduction techniques are applied. We also demonstrated that garnet AR layers can be grown at the magneto-optic film/substrate interface using a lithium yttrium molybdate flux. These AR films can be important in producing low insertion loss isolators in two ways. First, the film thickness can be cut in half with a structure which uses films on both sides of the substrate and this eases the film growth requirements and reduces defects in 2Bi/formula materials. The AR layer allows this double sided structure to be fabricated without internal backscattering or resonances from the substrate surfaces. Second, the AR layers can reduce forward insertion loss. In highly bismuth substituted films, the optical index gets large enough that 0.2dB reflection losses occur in a double sided structure. Finally, we have demonstrated that linear polarization rotation as a function of magnetic bias field is obtained in these types of films. This makes fabrication of devices which provide controlled variation of polarization easy to implement.

REFERENCES

1 E.C. Whitcomb et el, J. Appl. Phys. **49**(3), p1803-1805.
2 W.A.Bonner, Mat. Res. Bull. V12, p289-298.
3 G.L Nelson et el, J. Appl. Phys. **53**(3), p1687-1689.
4 V.J.Fratello et el, J. Cryst. Growth, V80 (1987) p26-32.
5 M. Kestigian et el, J. Crystal Growth, V42 (1977) p343-344.
6 Kobayshi et el, IEEE J. Quant. Elect. VQE-16,#1.
7 K.Tsushima, N. Koshizuka, Intermag 87.

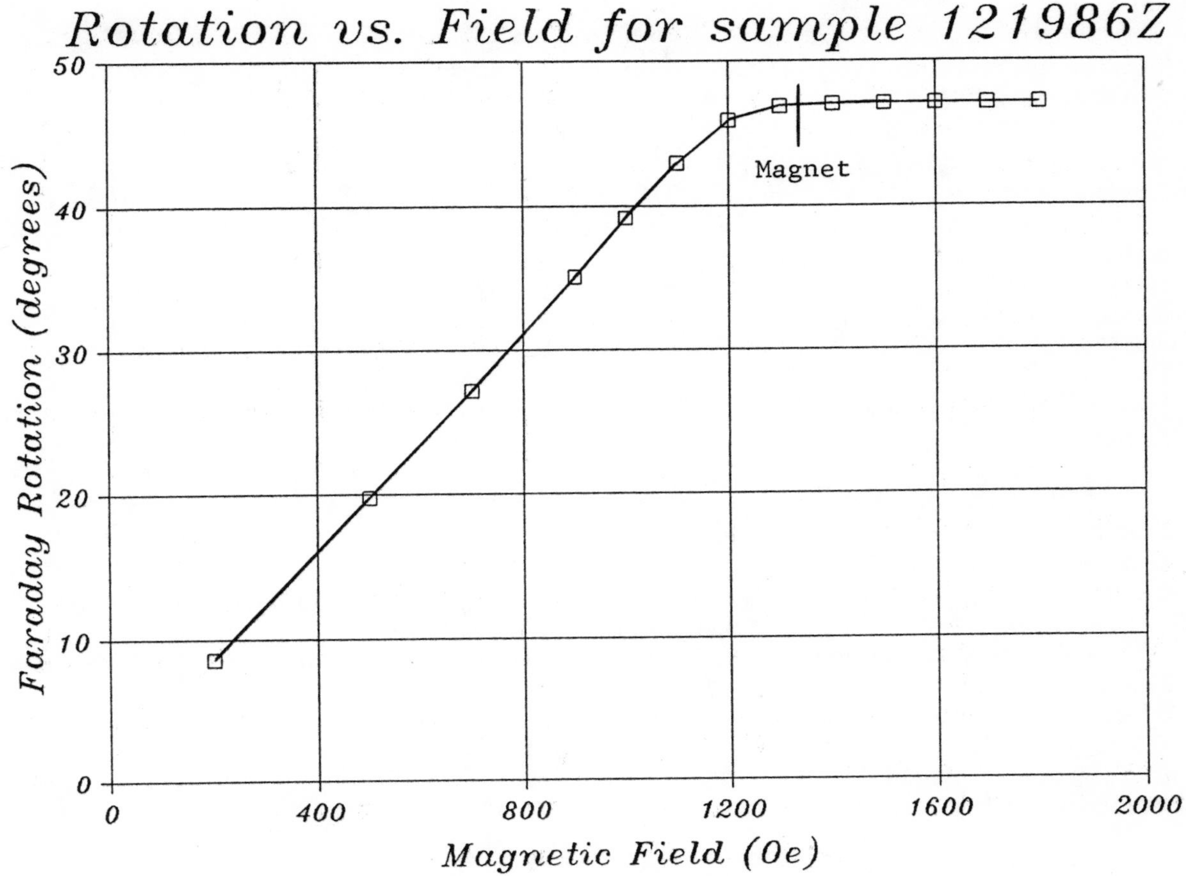

Figure 3

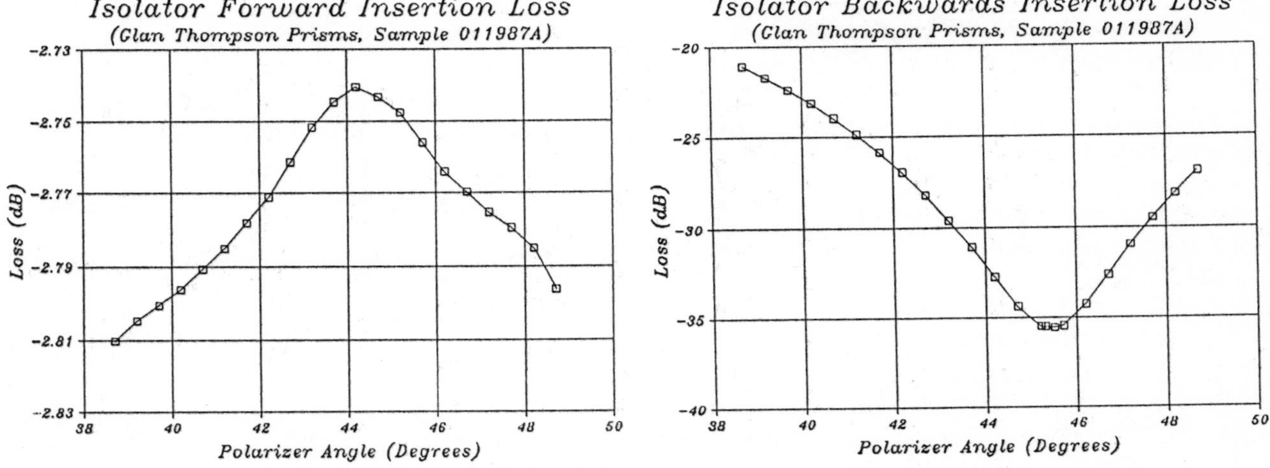

Figure 4

FRIDAY, JANUARY 29, 1988

MEETING ROOM 2-4-6

8:00 AM–10:00 AM

FAA1–8

CHEMICAL-BIOLOGICAL SENSORS

Annamaria Scheggi, Instituto de Ricerca Sulle Onde Electromagnetiche, Italy, *Presider*

DISTRIBUTED FLUORESCENCE OXYGEN SENSOR

R. A. Lieberman
L. L. Blyler
L. G. Cohen

AT&T Bell Laboratories
Murray Hill, New Jersey 07974

The evanescent field of light propagating in an optical fiber core extends into the cladding; this can cause the optical absorbance of the cladding to affect the transmission loss of the fiber itself. The fact that the evanescent fields of propagating modes in the fiber extend into the cladding also means that the inverse effect is possible: light generated in the cladding can be coupled to guided modes and transmitted over long distances. This work describes a fluorescence-based distributed oxygen sensor which makes use of this process.

The fiber used in these experiments was a plastic clad silica (PCS) structure, with a fluorescent sensor dye distributed uniformly throughout the cladding. Suprasil rod was drawn to a diameter of 125 μm and coated to a final diameter of 230 μm with a permeable dye-doped cladding in a procedure similar to that used by Blyler, Ferrara, and MacChesney, to create an absorbance-based distributed ammonia sensor.[1] The cladding material, 9,10-diphenylanthracene (9,10-D) in poly(dimethyl siloxane) (PDMS), was chosen for its known suitability as an oxygen sensing system,[2] and was heat-cured in place during the draw process.

After draw, fluorescence emission and excitation spectra for the fiber were measured, using the experimental arrangement shown in Figure 1. The results clearly demonstrated that some of the cladding fluorescence was guided by the fiber core. The fraction of cladding-generated fluorescence emission coupled to guided modes was measured by comparing the fiber results with the intensity of emission obtained from a 9,10-D/PDMS film excited with the same power density as the fiber. This fraction, $P_{core}/P_{tot} = 2 \times 10^{-4}$, is in agreement with numerical predictions.[3] The spectral attenuation ("loss") of the fiber was also measured and exhibits an absorbance maximum (0.38 dB/m) at 390 *nm*, corresponding with the excitation peak of 9,10-D, as well as secondary maxima at 720 and 875 *nm* due to C-H bonds in the silicone itself. The measured transmission loss at the peak emission wavelength of the 9,10-D/PDMS system (430 *nm*) was 0.29 dB/m for end coupled light.

The fluorescence of 9,10-D is quenched in the presence of oxygen, with the degree of quenching depending on oxygen concentration. Oxygen response of the sensor fiber was investigated using arc lamp excitation and lock-in detection, as illustrated in Figure 2. The fiber was coiled in a small chamber and illuminated from the side with 365 ± 10 *nm* light. Gas concentration in the chamber was varied by switching between different supplies containing calibrated concentrations of oxygen in nitrogen. The measured time constant for system response to changes in oxygen concentration is approximately 5 seconds, but this response is primarily determined by the time required to exchange atmospheres in the gas test chamber. It is therefore likely that the response time of the sensor is considerably shorter than the system response time, and may even be as short as 0.2 sec (the estimated time required for oxygen to diffuse from the surface of the cladding to the fiber core).

By measuring fluorescence output intensities for several different oxygen concentrations, it was possible to use the Stern-Volmer relation

$$\frac{I_0}{I} = 1 + \frac{P_{O_2}}{P'}$$

to calculate the half-quench pressure, $P' = 2.10 atm$, of the 9,10-D/PDMS system used in the fiber cladding.

The dependence of fluorescence output intensity on fiber length was investigated, and was found to be approximately linear, for lengths shorter than 10 *m*, as shown in Figure 3. This behavior is in agreement with theoretical predictions for the sections of fiber short compared to $10/\alpha$, where α is the fiber transmission loss measured at the fluorescence output wavelength.

Using data such as that shown in Figures 2 and 3 to calculate average signal-to-noise ratios, the oxygen sensitivity per unit length of the 9,10-D/PDMS clad fiber for the lock-in detection scheme was calculated to be approximately 10 $mmHg-m$ (e.g. a one-meter length of the fiber would have a minimum detectable oxygen partial pressure of about 10 $mmHg$). This number represents an upper bound on the minimum detectable oxygen pressure per unit length since no effort was made to reduce noise effects due to source fluctuations, bending variations, etc.

Photobleaching of the 9,10-D/PDMS cladding was also investigated by performing long-term illumination studies of the fiber in various gaseous atmospheres. Fluorescent output intensity was observed to follow the relation

$$I(t) = I_0 \times e^{-(t/\tau_b)}.$$

where the photobleaching decay constant, $\tau_b =$ 20 hours. The bleaching behavior was observed to have very little dependence on the atmosphere in which the samples were placed.

The spatially continuous nature of the oxygen sensor described in this report makes it suitable for a broad range of applications for which point sensors would be impractical. Distributed fiber optic sensors based on fluorescent claddings have quick response and high sensitivity and can be used in a mode which does not require a light source to be coupled to the end of the fiber. Fluorescent dyes have been developed which are sensitive to many chemical species, and should prove useful in the development of distributed sensors for species other than oxygen.

REFERENCES

1. Blyler, L. L., Ferrara, J. A., MacChesney, J. B., "A Plastic-Clad Silica Fiber Chemical Sensor for Ammonia," this conference (OFS '87).

2. Cox, M.E.; Dunn, B., "Detection of Oxygen by Fluorescence Quenching," *Appl. Opt.* **24**(14), P. 2114 (1985).

3. Marcuse, D.; *IEEE Journal of Lightwave Technology*, submitted July, 1987.

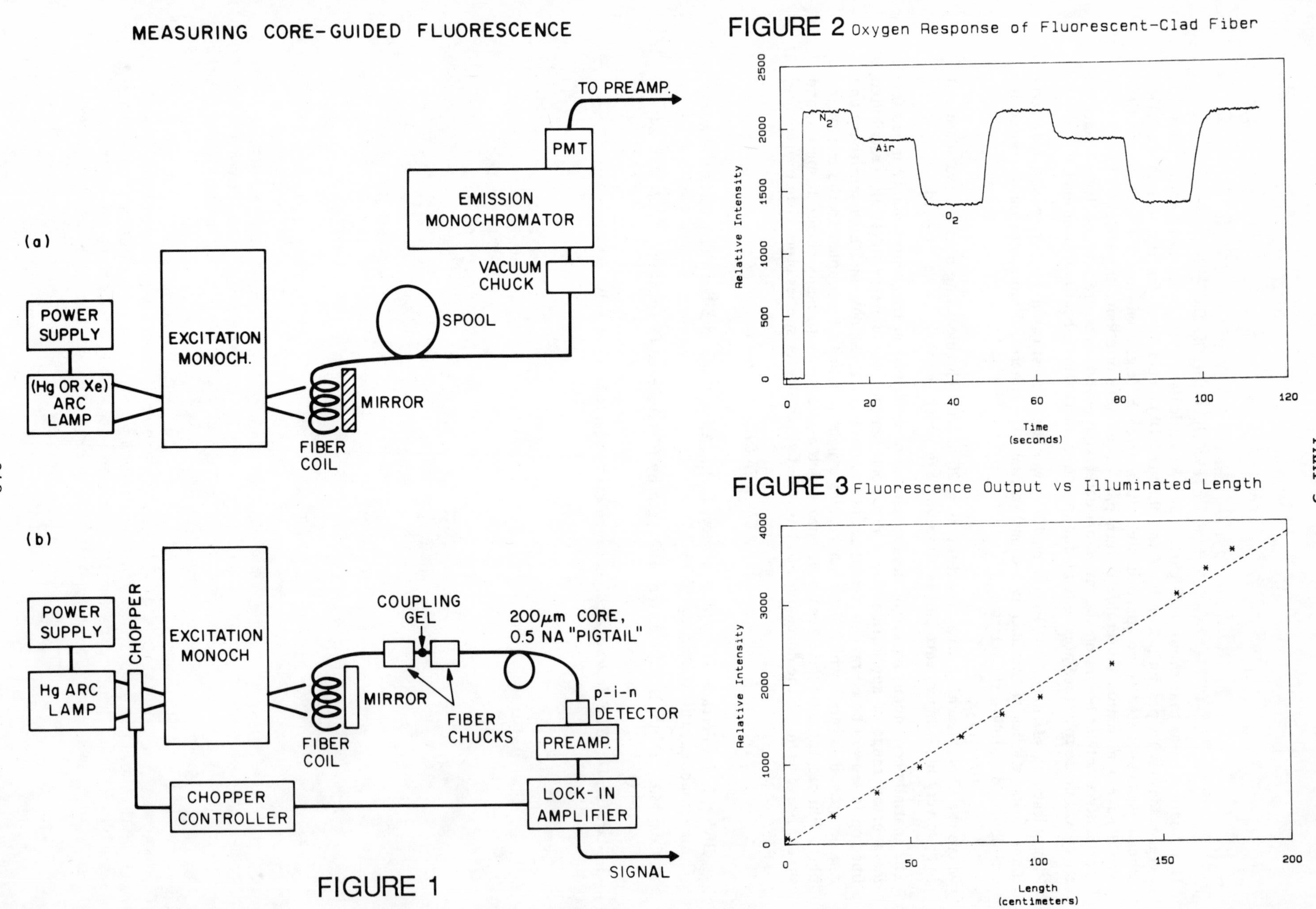

FIGURE 1

FIGURE 2 Oxygen Response of Fluorescent-Clad Fiber

FIGURE 3 Fluorescence Output vs Illuminated Length

THE DESIGN AND EVALUATION OF A REVERSIBLE FIBER OPTIC SENSOR FOR DETERMINATION OF OXYGEN

Marion R. Surgi

Allied-Signal Engineered Materials Research Center
50 E. Algonquin Road
Des Plaines, Illinois
60017-5016

SUMMARY

The study of fiber optic sensors for real time measurements is one of the fastest growing areas in analytical chemistry. With potential applications in biotechnology, blood chemistry, biofermentor control and anaerobic process control, measurement of gases has become increasingly important. In particular, oxygen has created a great deal of enthusiasm due to its biological importance and the number of biochemical reactions either using oxygen as a reactant or releasing it as a product. Currently, the most widely used method to determine oxygen concentration is the electrochemical sensor based on the Clark electrode.

Numerous problems associated with some electrochemical methods have prompted still other approaches which depend upon polarographic techniques.[2] Problems associated with required electrical connections, lack of reproducible results, the need of a steady-state supply of oxygen to the sensor surface, and the expense of producing a device which rigorously controls all the variables associated with polarographic or amperometric experiments has prompted the development of fiber optic oxygen sensors. These sensing devices depend upon light intensity measurements rather than current measurements to quantify oxygen. Oxygen is usually determined spectroscopically by indirect methods using a probe or indicator molecule.

Since oxygen is a voracious quencher of fluorescence, most approaches to date have utilized quenching of pyrene (or derivatives) fluorescence. Most sensors are limited to the use of these indicator molecules due to their inherently long fluorescence lifetimes. Such lifetimes are required if collisional deactivation is to compete favorably with light emitting processes. However, restrictions to using only one molecule severely limits the useful spectral regions. This is important since pyrene cannot be excited using a convenient laser source. The work presented here will discuss a systematic approach to induce room temperature phosphorescence allowing expansion of the number of indicator molecules available and increased sensitivity.

Only recently has attention been given to the interaction between indicator molecules and substrates for sensor applications.[3,4] In order to tailor a sensor for particular use, the interaction between the indicator molecule, the substrate and the membrane must be exploited. Consequently, the interaction of the silica used as substrate and the indicator molecule, N-methylacridone (NMA) is the most important interaction in this work.

Numerous investigations of aromatic carbonyl compounds deposited on silica reveal that the triplet states can be inverted if they are nearly isoenergetic.[5,6] McClure has shown[7] that spin-orbit coupling between $\pi\pi^*$ states of different multiplicities is negligible due to the vanishing one and two-center terms.

El-Sayed[8] has demonstrated that spin-orbit coupling between singlet and triplet nπ* and ππ* states is allowed. In other words, intersystem crossings from nπ* singlets to ππ* triplets or from ππ* singlets to nπ* triplets could be 3 to 4 orders of magnitude faster than crossings between nπ* singlets and triplets or ππ* singlets and triplets. If the triplet state is inverted, the lowest triplet usually becomes an ππ*. Hence, intersystem crossing from an nπ* singlet to a ππ* triplet is more probable.

We have identified an aromatic carbonyl, N-methylacridone (NMA), that when deposited on silica would be expected to undergo a similar triplet state inversion. However, previous work has demonstrated that emission from a triplet state occurs only at 77K. Previous authors have investigated room temperature phosphorescence,[9] but these systems lack the response time needed for real-time monitors and do not appear to have responses sufficient for viable sensors. More current interest involves application of this phenomenon to realtime monitoring.

Intersystem crossing is also aided by the presence of a heavy atom which provides a mechanism for spin-orbit coupling.[9] In some instances the heavy atom is covalently linked to the emitting species (bromonaphthalene) and in other instances the physical association of heavy atoms with the emitting molecule is sufficient to enhance intersystem crossing.[9]

Therefore, the solvent deposited with the lumophore can have a profound interaction with the lumophore. Although methylene chloride was selected as the solvent due to high NMA solubility, its heavy atom effects cannot be ignored. Current work involves the codeposition with NMA of other heavy atom containing solvents. Hence, the use of silica as the substrate and methylene chloride as the solvent should combine to increase intersystem crossing thereby enhancing the population of the triplet state. The evidence which indicates phosphorescence quenching by oxygen is summarized below.

1. The excited state lifetime (τ) has been experimentally estimated to be 10^{-4} sec. characteristic of a triplet emission or delayed fluorescence and is sufficiently long lived to be quantitatively quenched.

2. The extended calibration curve for NMA deposited on silica passes through a maximum and then decreases. Such behavior is indicative of solid surface phosphorescence.[1]

3. The emission from NMA deposited on silica is red shifted from NMA fluorescence emission by 40 nm. This shift is greater than could be accounted for by solvent effects alone. Maximum red shifts of 10 nm have been reported for acetophenone fluorescence when the solvent is changed from methylcyclohexane to methylcyclohexane/silica and 15 nm when the solvent is changed from methylcyclohexane to 85% H_3PO_4.[5]

4. A surface coverage of 0.23% of the total BET surface area is probably too low to promote the excimer formation necessary for delayed fluorescence emission or any type of triplet-triplet interaction.[11]

5. Use of CH_2Cl_2 as the solvent for deposition of NMA onto silica should enhance intersystem crossing to the triplet state by spin orbit coupling.

6. The interaction between the carbonyl n-electrons and NMA has been demonstrated. Experiments performed using NMA as the indicator molecule were repeated with the McMurry[12] dimerization product of two moles of NMA (10,10'-dimethyl-9,9'-biacridylidene). This molecule contains two N-methylacridine units connected by an ethylene linkage in the 9 and 9' positions. While the presence of silica had substantial effects on the emission of NMA, it had no detectable effects on the emission of the biacridylidene.

The sensor, constructed by inserting an excitation and collecting optical fiber into a teflon cell containing silica coated NMA, provided a second order response to oxygen concentrations from 0.5 to 100% O_2 (V:V) (Table and Figure). The relative standard error of the estimate about the non-linear regression line was 4% when one sensor was evaluated during replicate studies. However, when replicate sensors were fabricated, the relative standard error of the estimate increased to 10%.

References

1. Siggaard-Anderson, O. Blood Gases In "Fundamentals of Clinical Chemistry", 5th ed., N.W. Tietz, Ed., W. B. Saunders, Philadelphia, PA, 1976.

2. Conway, M., Dorbin, G.M., Ingram, D., McIntosh, N., Parker, D., Reynolds, E.O.R., and Stoutter, L.P., Pediatrics, 57, 244 (1976).

3. Martinho, J.M.G., and Winnik, M.A., J. Phys. Chem., 91, 3640 (1987).

4. Wyatt, W.A., Poirier, G.E., Bright, F.V., and Hieftie, G.M., Anal. Chem., 59, 572 (1987).

5. Lamola, A.A., J. Chem. Phys., 47, 4810 (1967).

6. Rusakowicz, R., Byers, G.W. and Leermakers, P.A., J. Am. Chem. Soc., 93, 3263 (1971).

7. McClure, D., J. Chem. Phys., 20, 682 (1952).

8. El-Sayed M.A., J. Chem. Phys., 38, 2834 (1963).

9. Lee, E.D., Weiner, T.C. and Seitz, W.R., Anal. Chem., 59, 279 (1986).

10. Burrell, G.J., and Hurtubise, R.J., Anal. Chem., 59, 965 (1987).

11. Bauer, R.K., Ware, W.R., and Wu, K.C., J. Phys. Chem., 86, 3781 (1982).

12. McMurry, I.E., Fleming, M.P., Kees, K.L., and Krepski, L.R., J. Org. Chem., 43, 3257 (1978).

FAA2-4

RESPONSE OF OXYGEN SENSOR

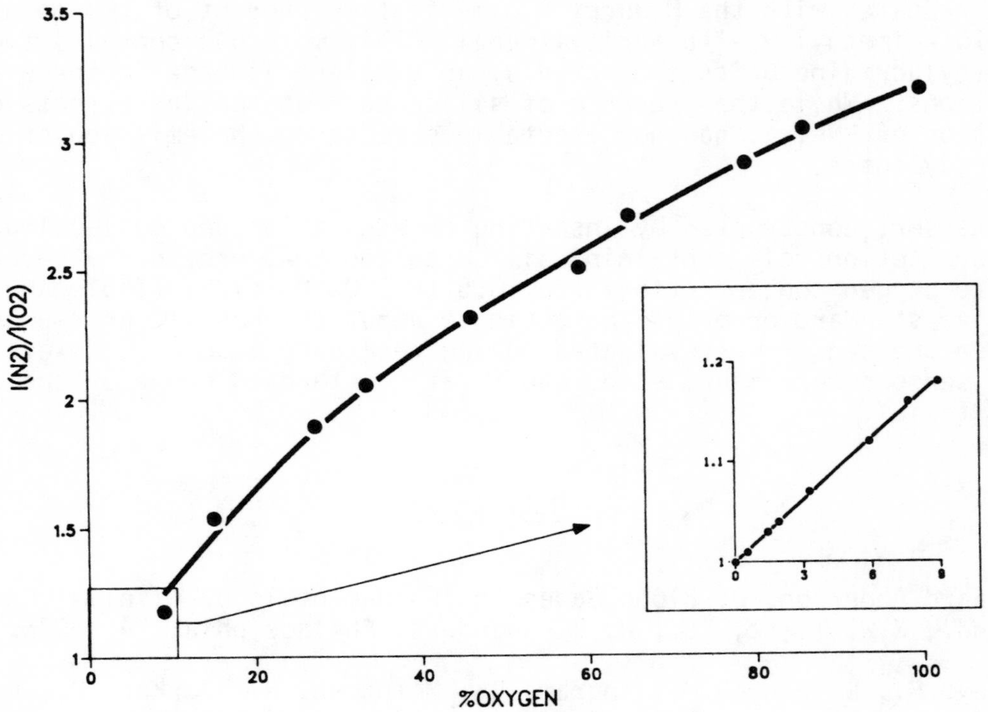

Oxygen Concentration[a] (% V:V) Experimental	Calculated[b]	I[N2]/I[O2][c] (460 ± 5 nm)
0	0.87	1.00
0.54	1.15	1.01
1.44	1.73	1.03
1.93	2.01	1.04
3.27	2.98	1.07
5.9	4.33	1.12
7.6	5.51	1.16
8.9	6.10	1.18
15.0	17.3	1.54
27.2	29.3	1.89
33.4	35.3	2.05
46.0	46.0	2.31
59	55	2.50
65	65	2.70
79	77	2.90
86	87	3.03
100	101	3.18

a) Correlation coefficient (R-squared) between experimental and calculated values: 0.997; 1σ = 1.8% O_2.

b) Calculated using the following equation to determine % O_2: $133 - [2.49 \times 10^4 - 7.52 \times 10^3 (I[N_2]/I[O_2])]^{1/2}$

c) $I[N_2]$ provided a signal of 270 namps (noise = 1 namp).

A new fiber-optic sensor for bile reflux

F.Baldini, R.Falciai, A.M. Scheggi
IROE-CNR, Via Panciatichi 64, 50127 Firenze, ITALY

P.Bechi
Patologia Speciale Chirurgica I dell'Università di Firenze
Viale Morgagni 85, 50134 Firenze, ITALY

INTRODUCTION

Absence of electromagnetic interference, electrical insulation, easy handling and miniaturization are well-known advantages of optical fiber sensors. In particular in the medical field [1-5], optical fiber sensors have undergone a great development, also because they can allow measurements of parameters otherwise undetectable.

This paper is dedicated to the description of a new optical fiber sensor for the detection of bile reflux.

Conventional techniques for testing acid gastro-esophageal reflux (GER) (from the stomach into the esophagus) [6] and entero-gastric reflux (from duodenum into stomach) are based on pH measurements of esophageal content and gastric juice. Such measurements are generally performed by positioning a pH glass electrode probe in the distal esophagus and in the gastric antrum respectively. Normally GER gives rise to a decrease of the almost neutral enviromental esophageal pH of about three or four units, while an entero-gastric reflux causes an increase of the acid basal gastric pH towards neutrality because of neutral or alkaline duodenal juice. However, some false positive results in detecting entero-gastric reflux may occur by pH-monitoring owing to the direct contact of the probe with the stomach neutral mucus [7]. Further, in some conditions (atrophic gastritis, partial gastrectomy, etc.), in which gastric pH is quasi neutral, entero-gastric reflux could be undetectable by pH monitoring [8]. It is important to observe that, in this case, should a GER occur, it could be undetectable because it is not acid enough to give rise to a relevant and clearly detectable esophageal pH modification.

The problem of detection of non-acid GER and entero-gastric reflux is not purely speculative; in fact the presence of bile and pancreatic juice in the reflux can enhance its harmful effect on the esophageal mucosa [9] . Moreover, while a clear anatomical and histological counterpart of entero-gastric reflux in the unoperated stomach has not yet been identified, erythema and foveolar hyperplasia [10] of the gastric mucosa have been shown to be reflux-related after partial gastrectomy. On the other hand reflux could be responsible of some "dyspeptic" syndromes of uncertain classification.

Both non acid- GER and entero-gastric reflux are partly made up of bile; for this reason a detection based on absorption properties of the biliar pigments (mainly bilirubin) can be a new method capable of revealing the reflux also in those conditions in which the pH electrode is unsuccessful.

EXPERIMENTAL SET UP AND RESULTS

As a preliminary study for the development of a fiber optic

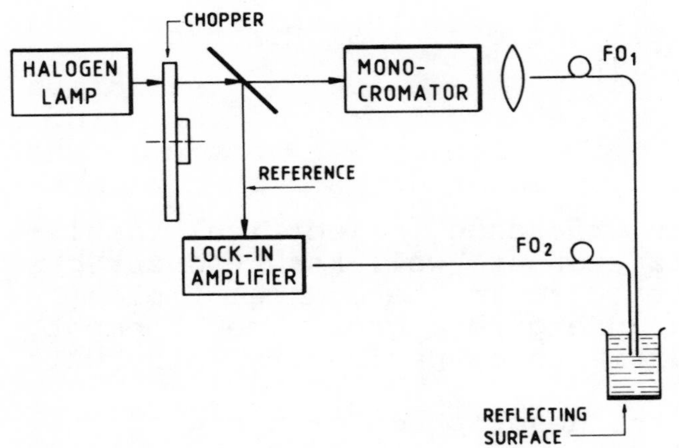

Fig.1
Experimental set up for absorption measurements

sensor, absorption properties of pure bilirubin and juice samples drawn from several patients were investigated in the 400-700 nm wavelength range by using the experimental set-up of Fig.1. The choppered light from an halogen lamp, wavelength-selected by a monochromator, is carried by an optical fiber (FO_1) to the sample under test, collected by a second optical fiber (FO_2) and conveyed to a lock-in amplifier. The measured transmittance spectra (normalized with respect to air) are shown in Fig.2: curve a) corresponds to a 10^{-4} M bilirubin solution in chloroform; curves b) and c) correspond to human bile and gastric juice from an ill patient respectively. The different intensity levels of the three curves are related to the different optical transparency of the three liquids where the bilirubin is dispersed. A well pronounced minimum at λ=460 nm is present in the three spectra while a second less pronounced minimum is observable in curves b) and c) at λ=650 nm, owing to the presence of another bile pigment (biliverdin). On the other hand, as measurements performed on pure gastric juice samples do not reveal any selective absorption in the considered band, the presence of bile reflux in the

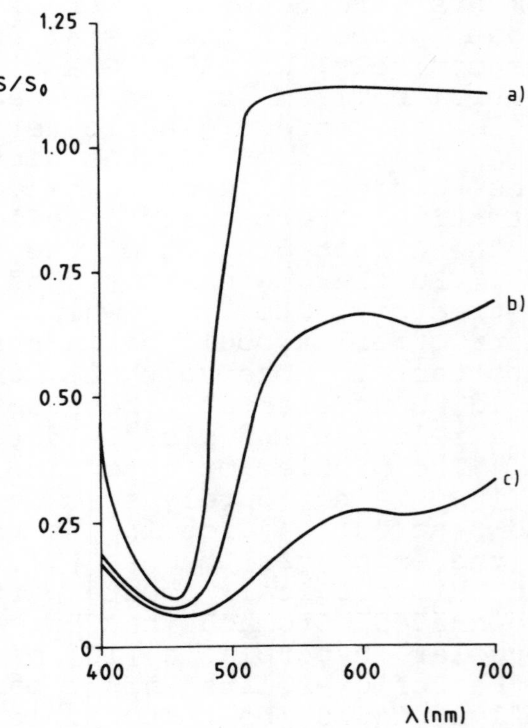

Fig.2
Normalized transmittance spectra of bilirubin a), human bile b) and gastric juice.

stomach can be detected by measuring the transmittance at λ= 460 nm and at a wavelength of high transmittance (for instance at λ= 600 nm). In the example of Fig.2 (curve c)), the relative variation between the transmittance at λ = 460 nm and λ = 600 nm is 70%.

On the basis of these results, an optical fiber sensor system has been designed and set up; the probe (Fig.3) is constituted by a stainless steel capillary (2 mm i.d.) with a reflecting bottom, and some holes drilled in the lateral surface to allow the flux of the juice sample under study; three PCS fibers (600 μm core diameter) are fixed into the capillary at a distance of about 1.5 mm from the specular bottom. The block diagram of the system is shown in Fig.4; optical fiber FO_1 carries the light from the source to the probe and FO_2 and FO_3 collect the reflected modulated light from the probe to the detectors. Two narrow-band filters select the light at λ= 460 nm and λ= 600 nm respectively and the two signals I and I_0 are amplified, filtered and ratioed, thus making the measurement also independent of possible source fluctuations.

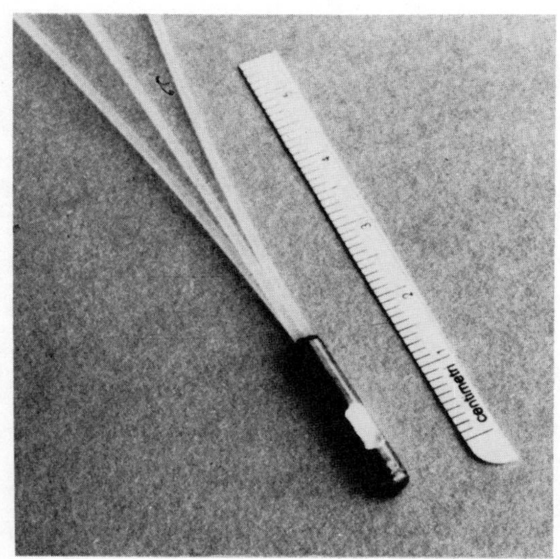

Fig.3 Miniaturized probe of the bile reflux sensor

In vitro tests on samples drawn from a number of patients have been carried out; after having dipped the probe in pure gastric juice, a sample drawn from an ill patient is added (in the ratio 1:10) in order to simulate the in vivo reflux conditions. Fig.5 shows the response curves of the sensor versus time for three typical samples obtained from patients with very high, medium and low enterogastric reflux respectively. Note that, although curve c) refers to a sample drawn from a patient with small reflux, it exhibits a variation of the signal perfectly observable of about 30%. This evidencies the potentiality of this technique for the detection of non-acid GER and entero-gastric reflux also in those conditions where pH electrode is unsuccessful. Finally the response time is very

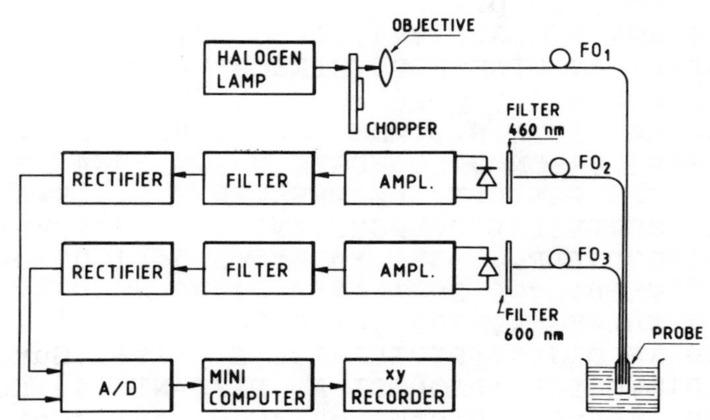

Fig.4 Block diagram of the sensor system.

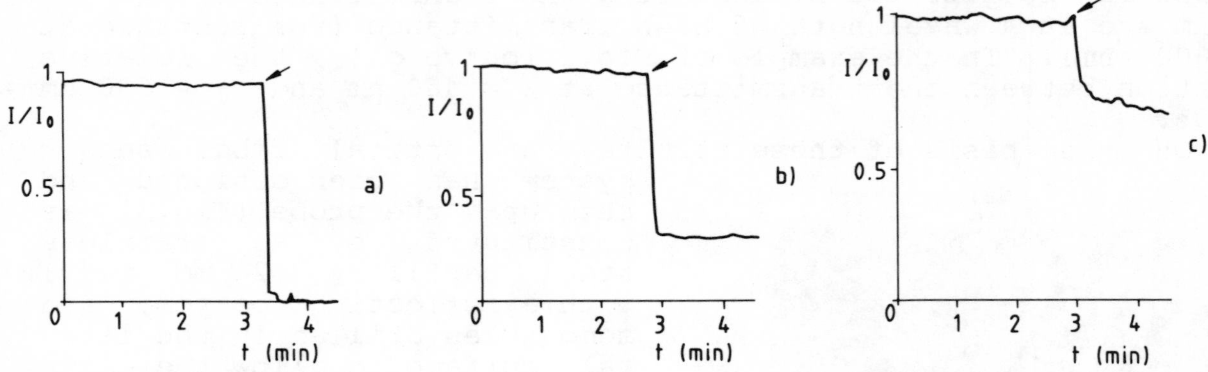

Fig.5 Response curves for three juice samples drawn from ill patients. Addition of ill patient gastric juice is performed in corresponding of the arrow.

fast, of the order of few seconds.

The system will be implemented by replacing the halogen lamp with two LEDs at the proper wavelenghts and using thinner fibers for further miniaturization of the probe.

However the probe is already sufficiently small to be used for in vivo measurements, assuring a better safety with less stress for the patient in comparison with the pH glass electrode, so far used in this type of illness.

REFERENCES

[1] SCHEGGI A.M., 1984, Proc. 2nd Intern.Conf. on Optical fiber sensors,p.13.
[2] VUREK G.G.,1984, Proc.SPIE, Vol.494,p.2.
[3] MILANOVICH F.P.,HIRSCHFELD T.B.,WANG F.T., 1984, Proc. SPIE, Vol. 494, p.18.
[4] HARMER A.H.,DAHNE C., 1985, Proc.SPIE, Vol.522,p.174.
[5] PETERSON J.I.,VUREK G.G., 1984, Science, 224, p.123.
[6] JOHNSON L.F.,DE MEESTER T.R.,1974,Am.J.Gastroenter.,63,p.325.
[7] CORTESINI C.,PUCCIANI F.,1984, Eur.Surg.Res,16,p.378.
[8] BECHI P.,1985, in Esophageal disorders: Pathophysiology and Therapy; De Meester T.R. and Skinner DB eds. Raven Press, New-York, 1985, p.635.
[9] SALO J.A.,KIVILAAKSO E., 1981,Surgery, 93, p.525.
[10]BECHI P.,AMOROSI A.,MAZZANTI R.,ROMAGNOLI P.,TONELLI L., 1987, Gastric histology and fasting bile reflux after partial gastrectomy, Gastroenterology (in press. Vol.94).

A FIBER - FABRY - PEROT MOTILITY SENSOR FOR THE MEASUREMENT OF PERISTALTIC MOTIONS IN THE UPPER GASTROINTESTINAL TRACT.

H. Wölfelschneider, R. Kist, Fraunhofer-Institut für Physikalische Messtechnik,
Heidenhofstr. 8, D 7800 Freiburg, Germany (FRG)

J. Schneider, H. Modler, Chirurgische Universitätsklinik, Robert-Koch-Strasse 40, D 3400 Göttingen, Germany (FRG)

Abstract

A fiber-optic sensor with a Fiber-Fabry-Perot (FFP) as sensing element has been developed and packaged to measure peristaltic motion in the oesophagus - stomach transition region. Clinical tests that have been performed with fifteen patients demonstrate that this new fiber-optic sensor does not exhibit problems such as sensor dislocation known from conventional sensor types, and that it is easily applicable and well tolerated by the patients.

Introduction

Perturbations of the duration, amplitude, and velocity of peristaltic motions of the tubular oesophagus as well as its closing segments (sphincters) constitute a pathological phenomenon of increasing importance. Therefore a suitable sensor is needed to diagnose pathological modifications of the peristaltic motion. Two types of conventional sensors exist that measure the contractions of the muscular oesophagus and its sphincters in terms of pressure variations. One type concerns directly operating electromagnetic pressure microtransducers (1), the other 4 side holes perfusion catheters (2) or 4 quadrant sleeves (3). Both sensor types suffer from dislocation problems that cause large scatter and poor reproducibility of the measured data. The fiber-optic motility sensor described in this paper does not show these problems and, in addition can easily be desinfected, and is well tolerated by the patients.

Sensor description

The basic concept of the motility sensor consists in using a low finesse Fiber-Fabry-Perot (FFP) resonator as a strain gauge to detect contraction and bending motions that characterize peristaltic motility. The principle of this FFP strain gauge has been described earlier (4), and it has been demonstrated to be a sensing element which is highly sensitive and tolerates large intensity variations (at least 15 dB) of the light flux in the powering and signal trans- mitting fibers. Fig. 1 shows the block diagram of the FFP motility sensor. The FFP is fabricated from a piece (10-cm long) of graded index fiber (30 µm core diameter) packaged within a silicone matrix of 8x8x40 mm^3. The single mode powering and the multimode (step index) signal transmission fibers are glued to the ends of the U-shaped FFP element. This choice of fibers simplifies considerably the fiber to FFP coupling as well as the manufacturing steps.

To get a link line insensitive output we use wavelength modulation of the laser diode (Hitachi HL7801) which provides the first and second derivative of the FFP signal. For a sinusoidal modulation of frequency ω the transmitted intensity can be expanded into the series:

$$I(\emptyset + \Delta\emptyset_o \sin\omega t) = I(\emptyset) + \frac{dI(\emptyset)}{d\emptyset} \cdot \Delta\emptyset_o \sin\omega t + \frac{d^2 I(\emptyset)}{4d0^2} \cdot \Delta\emptyset^2{}_o \cdot (1 - \cos 2\omega t) \quad (1)$$

where $\emptyset$ is the phase and $\Delta\emptyset_o$ the amplitude of the phase modulation due to wavelength modulation. According to equ. 1 synchronous demodulation on the fundamental and the second harmonic provides a signal proportional to the first and second derivative, respectively. These two signals are free of unwanted DC-components and contain complete information to reconstitute the optical path changes including their sign of the FFP. Since the counting is performed at the signal zero crossings, the method is highly independent on link line losses.

Fig. 2 shows the measuring arrangement for the FFP motility sensor along with the packaged sensor element, the power and signal fibers, the diode laser to fiber coupling module, as well as the receiver and signal evaluation unit.

Results

The FFP motility sensor has been tested on a mechanical motility simulator and in vivo on fifteen patients. Excursions of the motility simulator from 0.01 to 10 mm have been detected with several FFP-sensors independently in the frequency range 2 to 20 cycles per minute. The FFP - signals turned out to be highly consistent and reproducible. The transition zone of high peristaltic pressure between stomach and oesophagus could clearly be identified independently on the patient position (supine, sitting etc.). The measured curves are not affected by non-peristaltic disturbances such as respiration. Peristaltic contractions due to swallow action are clearly detected as shown by the example of Fig. 3.

Conclusions

The Fiber-Fabry-Perot resonator as a strain gauge sensing element has been applied to measure motility characteristics of the upper gastrointestinal tract. The packaged sensor has been clinically tested on a group of fifteen patients in order to measure peristaltic motions stimulated when pulling the sensor through the sphincter transition region and when setting a stimulus by swallowing action. The motility sensor produced signals of high signal to noise ratio and reproducibility and did not show disadvantages such as measuring point dislocation and susceptibility to mechanical and thermal perturbations that are typical for cenventional catheter sensors and microtransducers, respectively.

Acknowledgement

The authors are grateful to H. Höfflin for skilled technical assistance. Part of this work has been supported by the German Bundesministerium für Forschung und Technologie.

References

1. Winans CH S (1972), The pharyngoesophageal closure mechanism: a manometric study. Gastroenterology 63, 768 - 777.
2. Gauer OH, Gienapp E (1950), A miniature pressure - recording device. Science 112, 404 - 405.
3. Schneider J, to be published.
4. Kist R, Ramakrishnan S, Wölfelschneider H, The Fiber-Fabry-Perot and its applications as a fiber-optic sensor element, SPIE Vol. 586, Fiber Optic Sensors (1985), 126 - 133.

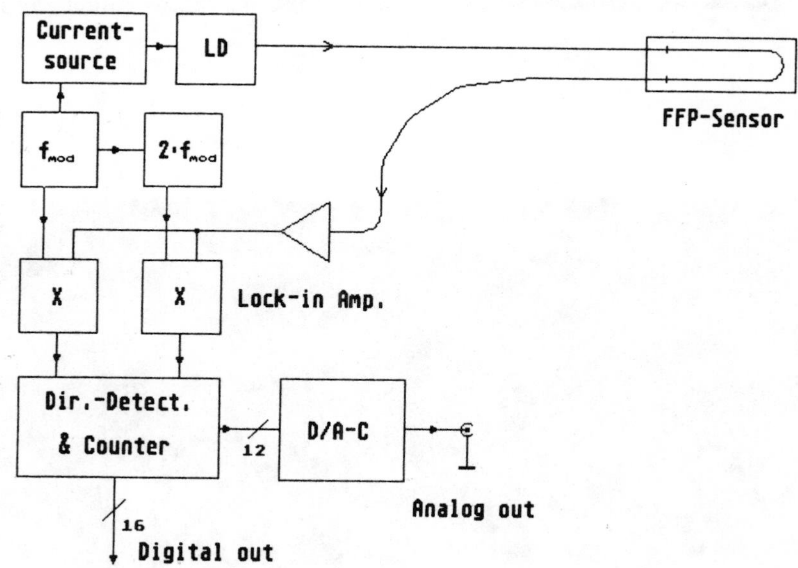

Fig. 1

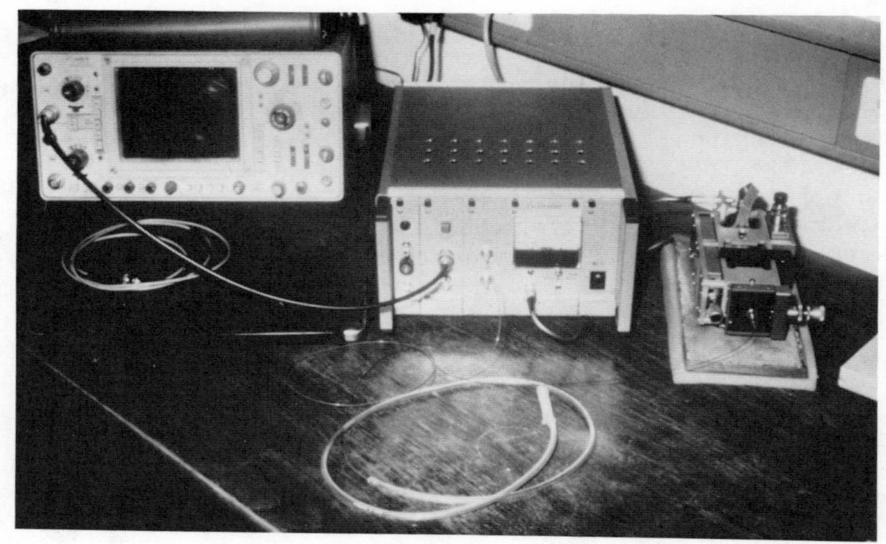

Fig. 2

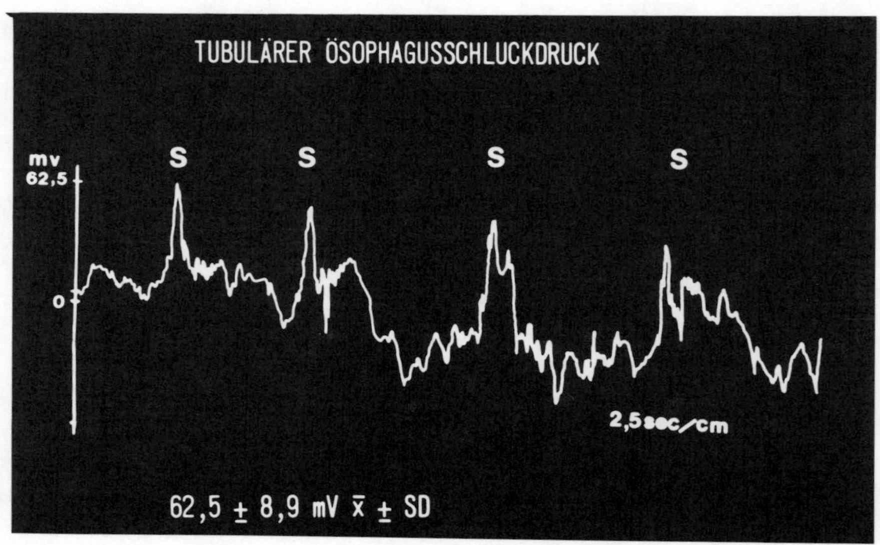
TUBULÄRER ÖSOPHAGUSSCHLUCKDRUCK
mv
62,5
0
S
S
S
S
2,5sec/cm
62,5 ± 8,9 mV x̄ ± SD

Fig. 3

WIDE RANGE SENSING OF LIQUID REFRACTIVE INDEX

Elric W. Saaski
Gordon L. Mitchell
James C. Hartl

Technology Dynamics, Inc./MetriCor, Inc.
18800 142nd Avenue N.E., Woodinville, WA 98072 U.S.A.
(206) 483-5577

Traditionally, the measurement of liquid refractive index has involved relatively bulky equipment such as the Abbe refractometer and total internal reflection to define refractive index to 10^{-4} refractive index units. This paper describes a new refractive index sensor which is approximately .2x.5x.5 mm in size and can be used with an optical fiber to sense the refractive index of liquids using less than 0.01 microliter samples.

The sensor is usable for virtually any liquid compatible with transparent refractory glasses or ceramics and silicon, and, can be used for monitoring liquids with refractive indexes from 1.0 to over 2.0. Limitations do, however, exist relative to the resolution obtainable for a particular refractive index span. Resolution comparable to an Abbe refractometer is possible when the total span is less than 0.10 refractive index units. Larger spans result in less resolution - in general, a resolution of .1% of full scale is obtainable.

The sensor uses a cavity resonator microshift technique to obtain an inherently monotonic optical response over a wide range of refractive index and is constructed as shown in Figure 1.

The sensor basically consists of a precision etched channel nominally 1 micron deep that has been formed in a glass or transparent refractory substrate and covered over with a polished silicon slab. In operation, the channel fills with the liquid to be analyzed primarily through the combined processes of capillary action and mass diffusion. The channel width is such that the optical fiber core shown is completely within its bounds. The optical fiber connects into an electro-optic system as shown in Figure 2. This system injects light from an infrared LED into the fiber, and contains a dichroic 45° mirror and two photodiodes to analyze return light from the sensor.

The sensor's principle of operation is as follows: Light from the LED passes through the fiber and transparent substrate and is reflected by the substrate/liquid/silicon structure. This structure forms a Fabry-Perot etalon which modulates reflected light depending on spacing between the reflector and substrate and in particular on the refractive index of the material contained in the channel.

The overall reflectance of this sensor for an 828 nm LED is illustrated in Figure 3. Here, it can be seen that the reflectance as a function of refractive index is a relatively nonlinear function. However, by splitting the return spectrum and ratioing integrated light intensities in the short and long wavelength wavebands as in Figure 3, the system can be linearized significantly. Figure 4 presents the calculated ratio of short versus long wavelength photocurrent as a function of refractive index for the channel used in the Figure 3 calculation. Figure 5 shows typical data for a 1.05 micron deep channel as tested with 11 fluids having refractive indexes ranging from 1.30 to 1.80. Note the monotonic and reasonably linear photocurrent ratio response for the refractive index range 1.30 to 1.60.

This new refractometer has been integrated with a microprocessor-based instrument to provide industrial refractive index measurement capability for an on-line processor. It has quick response and provides the capability to do refractive index measurements on-line with very small samples.

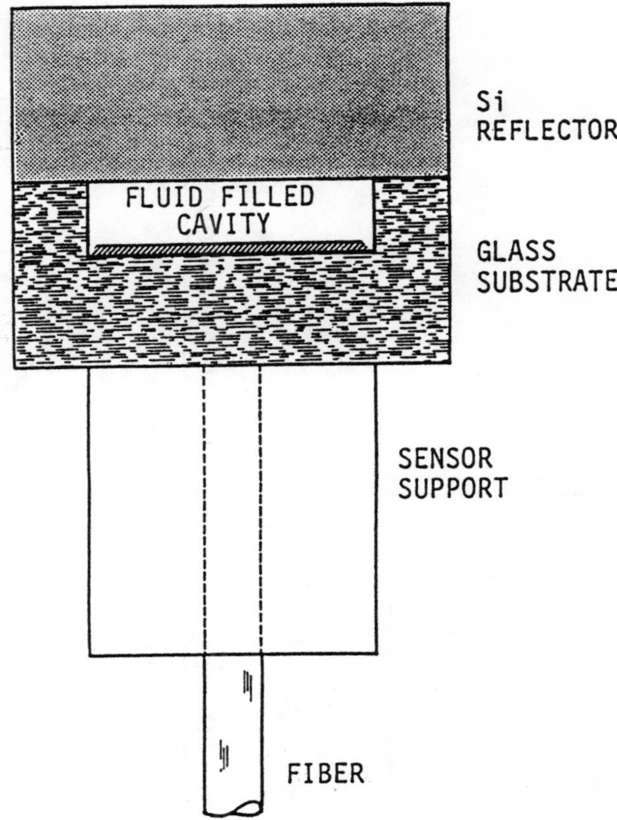

Figure 1. Refractive index sensor construction.

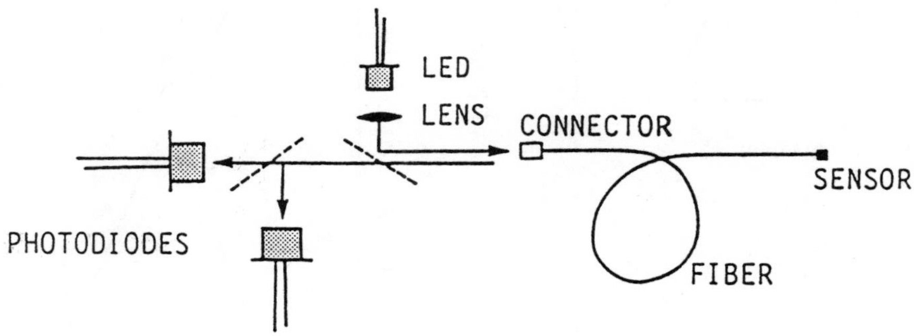

Figure 2. Electro-optic system used to measure refractive index.

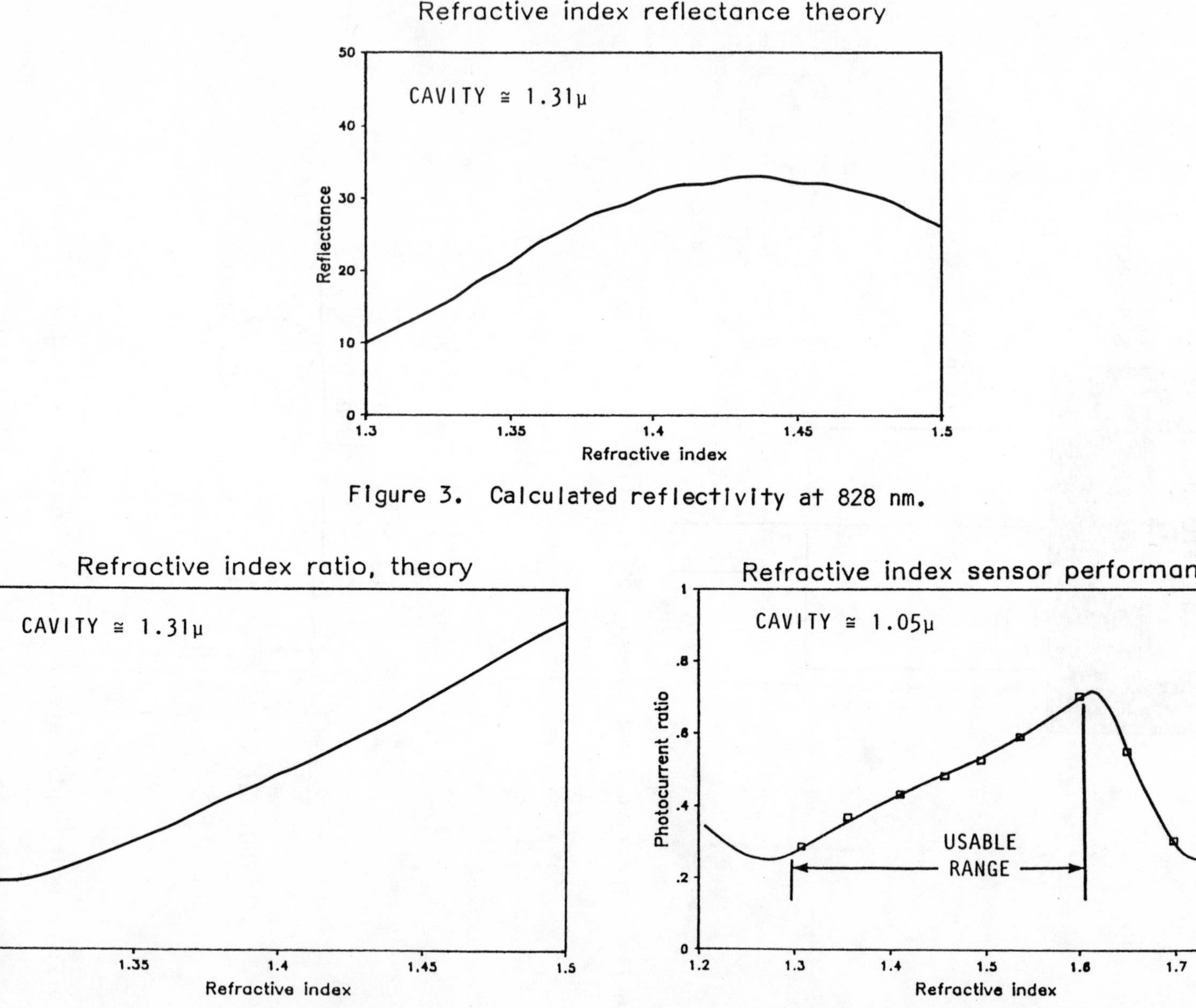

Figure 3. Calculated reflectivity at 828 nm.

Figure 4. Calculated photocurrent ratio.

Figure 5. Photocurrent ratio versus refractive index.

SUMMARY OF

REMOTELY SENSING MOLECULAR SPECTRA WITH FLUORIDE FIBER[1]

by

Pierre de Rochemont, Allen Bonde, Jr.[2], Anthony Boniface[3], Philip Levin

SpecTran Corporation
50 Hall Road
Sturbridge, MA 01566

and

Major Daniel Rapp
USMC Development Center
Quantico, VA

Fourier Transform Infrared Spectroscopy (FTIRS) is a rapid, inexpensive, and conclusive technique which can be used by the analytic chemist or material scientist for molecular identification of solids, liquids and gases by relating spectral absorptance to electronic, rotational and vibrational energy state transitions in the material under study (1). Until recently, studies of this kind have been restricted to laboratory conditions. The development of the compositional family of glass fluorides of zirconium, barium, lanthanum, aluminum, and sodium (abbreviated ZBLAN) for long length optical fiber telecommunications can now provide a mid-infrared transmitting optical link whereby samples, remotely positioned from the interferometer, can be sensed and analyzed. A mid-infrared optical window is of interest because this is the "signature region" for optical absorption spectra of many chemical species. This paper reports on advances in this technology utilizing direct chemical absorption and evanescent wave absorption as sensor probes.

The optical configuration of this remote sensing scheme is diagramed in Figure 1. Broadband blackbody radiation is launched into a Michelson interferometer where a potassium-bormide (KBr) beamsplitter redirects the beam along two separate arms. Laser interferometry is used to calibrate modifications to the optical path length in one of the arms, while the optical path of the other arm remains fixed. Both arms of the split beam are returned to the beamsplitter where they recombine with a phase relationship proportional to the difference in the optical path lengths, and then pass onto the optical fiber launching stage. An optical fiber link is again established between the sample stage and a narrow band, high sensitivity indium antimonide (InSb) photovoltaic junction detecting the resultant interferogram to be Fourier analyzed by the CPU. As a single-beam spectrum is employed in this work, a reference spectrum is acquired prior to exposing the sensor probe to the sample.

---

1 This research has been funded by the USMC through the SBIR program (Contract #M-00027-87-C-0016).

2 Cooperative Education Program: Worcester Polytechnic Institute, Department of Electrical Engineering, 100 Institute Rd., Worcester, MA 01609.

3 Cooperative Education Program: University of Massachusetts, Amherst, Department of Electrical Engineering, Marston Hall, Amherst, MA 01003.

FTIRS technology has already been applied to ZBLAN systems to measure spectral attenuation in optical fiber (2). The spectral window of this sensing scheme, with the sensor probes remotely positioned by two meters is depicted in Figure 2. All of the fibers used in this apparatus have been protected with a thin film ceramic to prevent atmospheric attack on the exposed glass surfaces.

Results for two sensing probes will be reported in this presentation. One probe uses a direct absorption cell where light, fiber-guided from the interferometer, is launched directly into the sample medium and partially absorbed. The "frequency modulated" light is then recollected and fiber-piped to the detector. Results obtained using this technique have demonstrated feasibility in applying this technology as a remote sensor of solids, liquids and gases. Figure 3 demonstrates a very strong correlation in the principal absorption bands of polystyrene thin films measured by conventional FTIRS to those sensed remotely through a 2 meter length of fiber; and, Figure 4 reveals that acetone vapor may also be remotely detected and its spectrum correlated with a direct scan of the vapor.

Progress made with evanescent wave absorption (3) will also be discussed. In the evanescent absorption approach, the active portion of the sensor stage is a completely clad-free, hermetically coated length of fluoride fiber where the atmosphere surrounding the fiber, air plus the sample, serves as a lossy optical cladding absorptive at frequencies identifying the sample. Cladded fluoride fibers, with cores matching the outer diameter of the sensing fiber, provide the optical link between the sensor stage and the interferometer. Care has been taken to select a hermetic coating for this application which is transmissive over the infrared frequencies of interest (4).

Particular emphasis will be placed on the sensing capabilities of each method with respect to critical parameters of the device, i.e., remote lengths, sample densities, optical coupling schemes, and in the case of the evanescent sensor, probe design sensitivities.

The accuracy and the versatility of FTIRS as a detection system is already recognized across many lines of technological innovation. The development of the ZBLAN family of glasses and their intrinsic ability to transmit broadly over the mid-infrared offers another means by which FTIRS can be applied rather simply. These methods used to remotely detect molecular spectra, in conjunction with efforts ongoing to improve upon state of the art optical fiber and sensor capabilities, lend much promise to this technology.

(1) "Fourier Transform Infrared Spectroscopy", Theophanides, T., ed., D. Reidal Publishing Co., Dordrecht, Holland, 1984.

(2) Ohishi, Y. & Sakaguchi, S., "Spectral Attenuation Measurements for Fluoride Glass Single-Mode Fibers by Fourier Transform Techniques", Electronic Letters, vol. 23, no. 6, 272 (1987).

(3) Injeyan, H., Stasfudd, O.M., & Alexopoulos, N.G., "Light Amplification by Evanescent Wave Coupling in a Multimode Fiber.", Applied Optics, 21, 1928 (1982).

(4) Private Communication, L. Vacha, SpecTran Corporation.

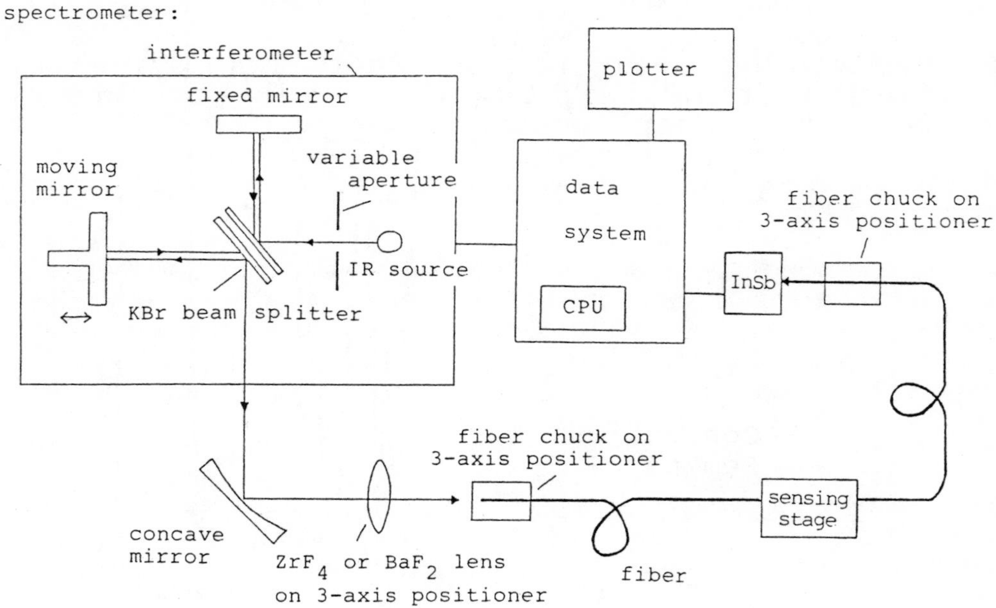

Figure 1: Chemical Sensing Scheme

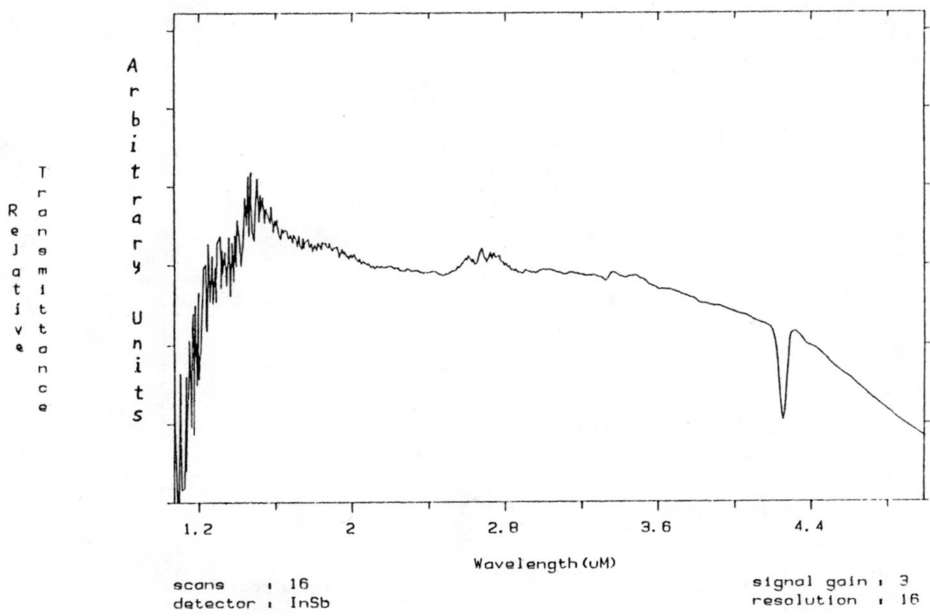

Figure 2: Relative Spectral Response of the Sensor, primarily limited by the InSb detector.

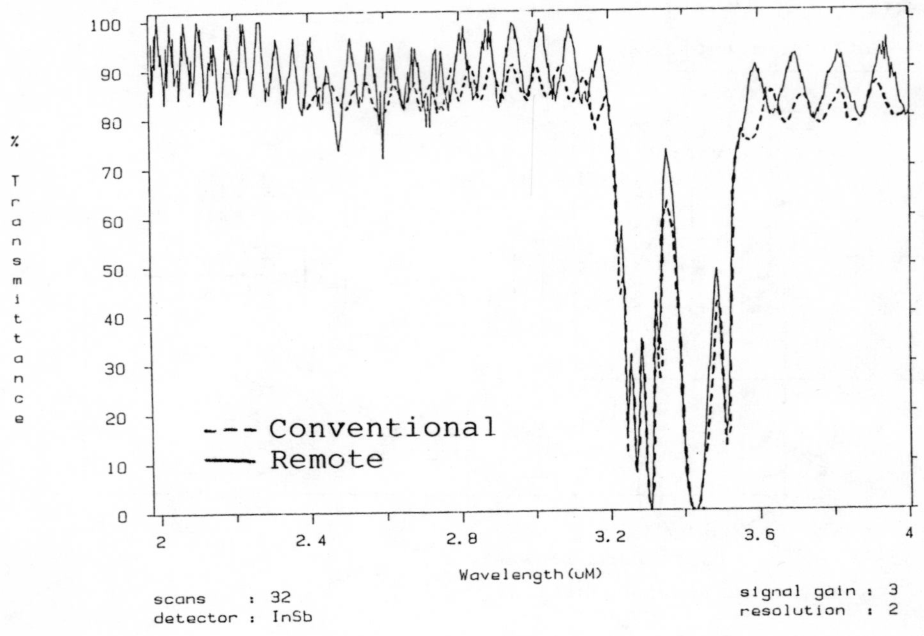

Figure 3: Polystyrene Sample;
Conventional FTIRS vs. Remotely Sensed

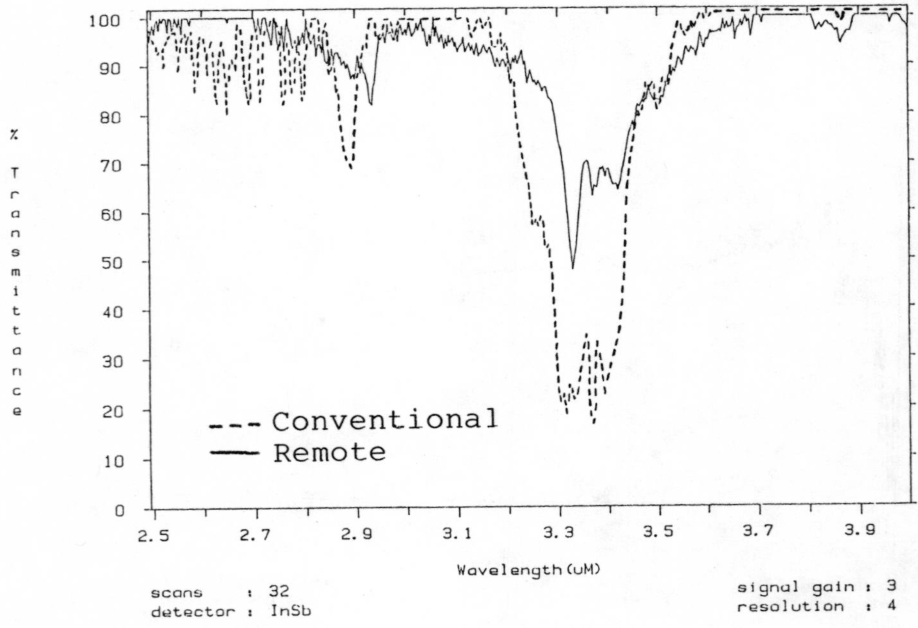

Figure 4: Acetone Vapor;
Conventional FTIRS vs. Remotely Sensed

A PLASTIC-CLAD SILICA FIBER CHEMICAL SENSOR FOR AMMONIA

L. L. Blyler, Jr.
J. A. Ferrara
J. B. MacChesney

AT&T Bell Laboratories
600 Mountain Avenue
Murray Hill, New Jersey 07974

Plastic-clad silica (PCS) fibers having a rubbery silicone cladding offer a distinct advantage for chemical sensing applications in that small molecule chemical species will diffuse through the cladding very rapidly. If a dye is incorporated in the silicone which is altered in color by a chemical of interest, the dye will be affected throughout the silicone volume. Further, because the evanescent tail of the light energy propagating in the fiber core extends into the cladding where it interacts with the dye, the absorption loss spectrum of the fiber is altered by the presence of the dye and by any color changes which it undergoes. The fiber thus represents a distributed chemical sensor in that absorption loss changes which occur anywhere along its length may be detected.

Oxazine perchlorate has been used to detect ammonia by earlier workers (1). It is blue in acidic and neutral media with a peak absorption at a wavelength of 665 nm, and red in basic environment with a peak absorption at 534 nm.

In this work a dye-doped PCS fiber was fabricated for the detection of ammonia. The dye was dissolved into a photocurable silicone acrylate formulation ($n_d = 1.42$) used to clad the silica core.

Several hundred meters of fiber were drawn from a high purity synthetic silica (Suprasil 2) preform rod and clad with the dye-doped, UV-curable silicone acrylate. A test length of this fiber on a plastic spool was placed in a 1-gallon container which could be purged with anhydrous ammonia, and/or air in any desired mix ratio. The ends of the fiber were led out of the container to a spectral loss test set.

Figure 1 shows the spectral loss characteristics of the dye-doped silicone acrylate-clad fiber in air and in 100% ammonia. The principal difference in the loss curves is the dye-related absorption peak at 665 nm which disappears in the ammonia atmosphere.

The fiber loss at 665 nm changes from a value of 450 dB/km in air to 90 dB/km in 100% ammonia.

With the assumption that the normalized transmission loss change at 665 nm is linearly related to normalized ammonia concentration, the time-dependent response of the PCS ammonia sensor was analyzed assuming molecular (Fickian) diffusion through the rubbery silicone to be the controlling mechanism (2). The logarithm of the normalized loss change is plotted against linear time in Figure 2 for both sudden ammonia exposure after equilibration in air and sudden air exposure after equilibration in 100% ammonia. For the former case there is an initial time delay of 7 seconds, principally related to container fill time, followed by a linear region from 7 to 25 seconds. From the slope of this curve the diffusivity of ammonia in the rubbery polymer cladding is determined to be $2x10^{-6}$ cm^2/sec. The sudden air exposure response of the sensor is considerably slower than the ammonia exposure response and reversibility is incomplete, implying an interfering chemical interaction between ammonia and a component of the silicone acrylate cladding. The equilibrium response of the sensor to changes in ammonia concentration is depicted in Figure 3. The concentration dependence is non-linear with highest sensitivity in the low concentration range.

The dye-doped PCS fiber has demonstrated several advantages for chemical detection, which include high sensitivity, rapid response and distributed sensing. Deficiencies observed in this work, such as incomplete reversibility, non-linear concentration response and non-specificity, may be addressed through appropriate dye and cladding polymer formulation.

REFERENCES

1. J. F. Giuliani, H. Wohltjen, and N. L. Jarvis, Optics Lett., 8, 54 (1983).
2. J. Crank, The Mathematics of Diffusion, Oxford University Press, London, U.K., 1975, Chapter 5.

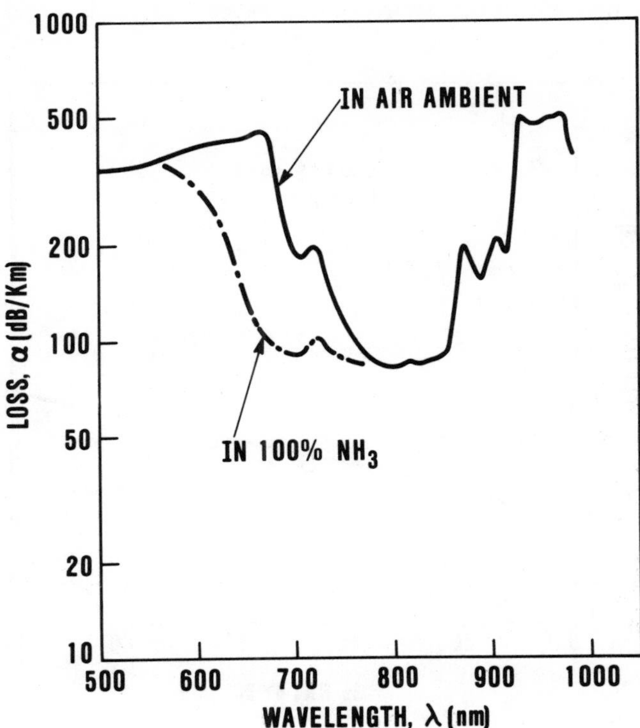

Figure 1: Spectral losses of PCS fiber ammonia sensor in air and in 100% ammonia.

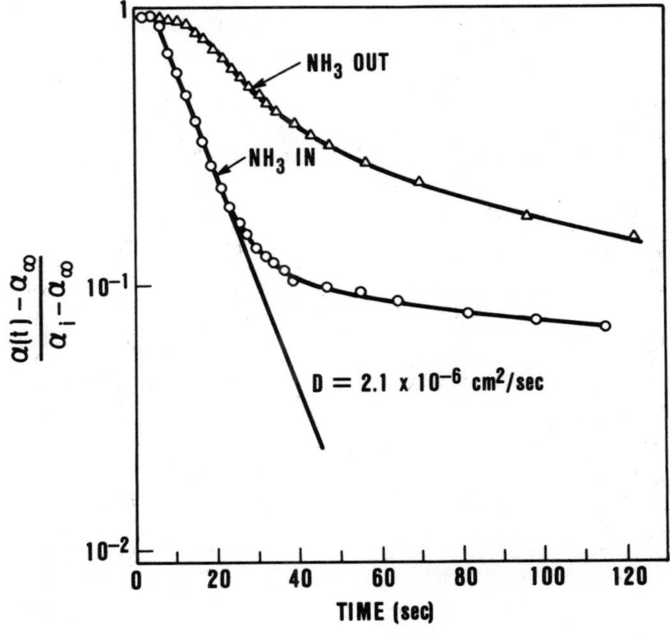

Figure 2: Time-dependent loss changes of 665 nm for PCS ammonia sensor.

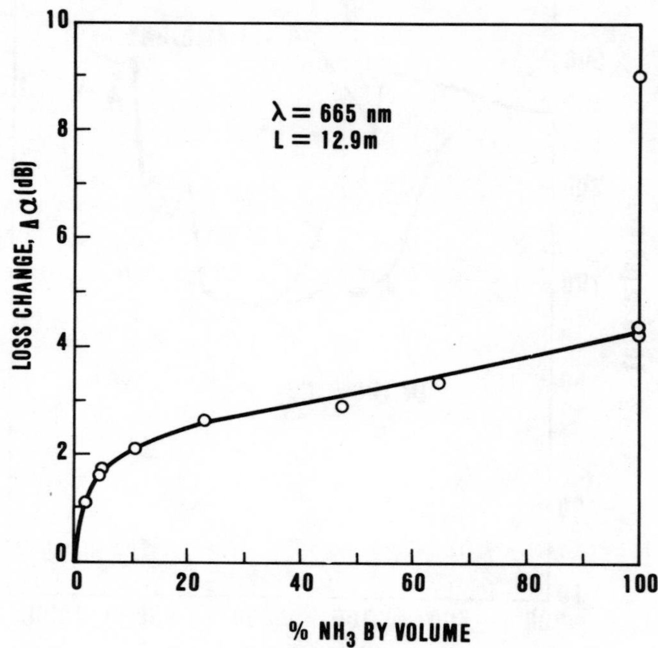

Figure 3: Concentration-dependent loss changes for PCS ammonia sensor.

POROUS FIBER OPTIC FOR A HIGH SENSITIVITY HUMIDITY SENSOR

Mahmoud R. Shahriari, George H. Sigel, Jr., and Quan Zhou, Rutgers University
Fiber Optic Materials Research Program, P.O. Box 909, Piscataway, NJ 08854

SUMMARY

Several approaches have been previously investigated which utilize fiber optic sensors to determine the relative humidity of air. Such devices typically have the advantage of higher sensitivity and quantitative precision compared to traditional chemical indicators. For example, the use of an optical fiber evanescent sensor for humidity measurements has been described by Russell and Fletcher[1]. In their device, a moisture sensitive cobalt chloride/gelatin film is immobilized on a 12cm long silica optical fiber as the humidity probe. Ballantine and Wohltjen described an optical waveguide humidity sensor that employed the same colorimetric reagent/polymer system on a glass capillary[2]. Zhu and Hieftje utilized a fluorescent fiber optic sensor for atmospheric humidity[3] based on a dye entrapped within a polymer matrix. However, all of these sensors can only determine the relative humidity above 40% because of sensitivity limitations.

In the present work, a new, high sensitivity, reversible porous fiber optic sensor for humidity measurement has been developed. Porous glass waveguides ranging from 150 to 300 microns were made by sequential heat treatment, phase separation and chemical leaching of a borosilicate glass fiber. This kind of waveguide has been characterized by scanning electron microscopy, BET measurements and mercury porosimetry. The results show that the porous fiber has an interconnective structure, a high surface area of 200 M^2/g and an average pore size of 1000 A. The humidity sensor is based on in-line optical absorption in porous glass fiber core and has much higher sensitivity than previous sensors based on either evanescent surface absorption or fluorescence.

Shahriari et al - "Porous fiber....."

Cobalt (II) chloride is used as the colorimetric reagent permeated into a 0.5cm length porous optical waveguide. The anhydrous cobalt chloride exhibits a strong absorption peak between 600-750nm. When water vapor diffuses into the porous glass fiber and hydrates the salt, the absorption peak shifts to 500nm. The experimental data show that the optical intensity of the absorption peak is proportional to the water vapor concentration. The transmitted intensity at 690nm is measured through the optical fiber to determine the humidity of air. Figure la shows the experimental arrangement used for this sensor with an expanded view of the humidity chamber which houses the porous glass fiber given in Figure 1b.

Figure 2 depicts a series of typical time response curves for the sensor which was exposed to a number of step changes in humidity. When the moisture is admitted into the chamber, the device responds rapidly and after a few minutes the output signal approaches a stable value. Conversely, when the flow of water vapor is stopped and only dry air is admitted into the gas chamber, the signal rapidly decays back to the baseline observed before the admittance of water vapor. Reduction in fiber diameter, increase in pore size or elevation of temperature will result in faster response times. Room temperature calibration curves for the device are shown in Figure 3 in which the changes in optical transmittance at 690nm are plotted as a function of the changes of relative humidity. Figure 3 also demonstrates that the sensitivity range of relative humidity can be changed by permeating the porous fiber with different concentrations of cobalt chloride. By using diluted cobalt chloride solution, the calibration curve A was obtained. The minimum detected relative humidity is about 0.5% RH at 25°C. The sensor can detect the higher level relative humidity when treated with a proportionately higher concentration of cobalt chloride. This behavior is completely reversible. Repeated cycling of the sensor over several days showed a reproducibility between optical intensity and relative humidity of 5%.

The porous glass fiber sensor described here for humidity monitoring is expected to see much broader application for both chemical and biochemical sensing. By selective activation of the glass surface with different chemical species, it is possible to target specific gases and liquids for on-line sensing. The technique lends itself to multiple sensing of either different chemical species or monitoring of the same species at various locations. In the latter case, a high resolution OTDR can be employed to measure extended system response. Porous glass fiber sensors appear particularly attractive for real-time, environmental monitoring of air and water pollution.

In summary, a new high sensitivity fiber optic moisture sensor has been developed. The device utilizes a pretreated, high surface area porous glass structure which allows either gas or fluid to permeate through the fiber core. This has permitted relative humidity measurements at much lower levels than any previously reported fiber devices along with excellent response times and full reversibility.

Shahriari et al - "Porous fiber....."

REFERENCES

1. A.P. Russell and K.S. Fletcher. Anal. Chim. Acta. 170, 209 (1985).

2. David S. Ballantine and Hank Wohltjen. Anal. Chem. 58, 2883 (1986).

3. Chu Zhu and Gary M. Hieftje. Abstract 606, paper presented at the Pittsburgh Conference and Exposition on Analytical Chemistry and Applied Spectroscopy, Atlantic City, NJ, 1987.

Shahriari et al - "Porous fiber....."

Figure 1a. Schematic diagram of the experimental arrangement of the humidity sensor: a) quartz halogen lamp, b) chopper, c) monochromator, d) humidity chamber, e) photodiode, f) lock-in amplifier, g) chart recorder or computer.

Figure 1b. Expanded view of humidity chamber illustrating porous glass fiber.

Figure 2. The typical response curves of humidity sensor at $\lambda = 690$nm to a continuous flow of moist air. The flow rate is 1 liter/minute at 25C. A) 8.2% RH, B) 10.4% RH, C) 12.1% RH, E) 14.2% RH, F) 15.7% RH.

Figure 3. Humidity calibration curves for porous fibers permeated with different cobalt chloride concentrations: A) 10 mg/ml, B) 139 mg/ml, C) 213 mg/ml.

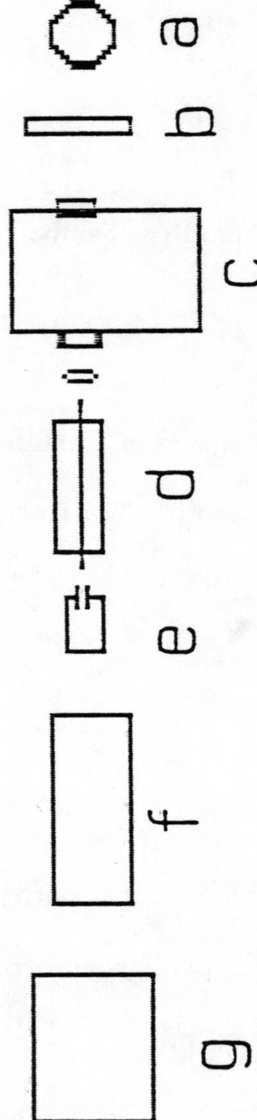
a
b
c
d
e
f
g

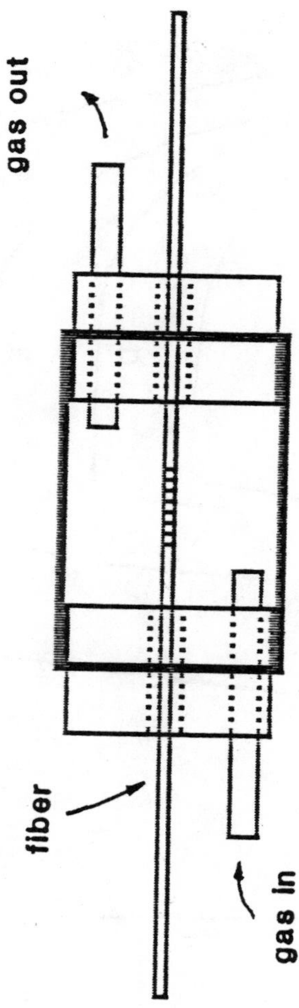
gas out
fiber
gas in

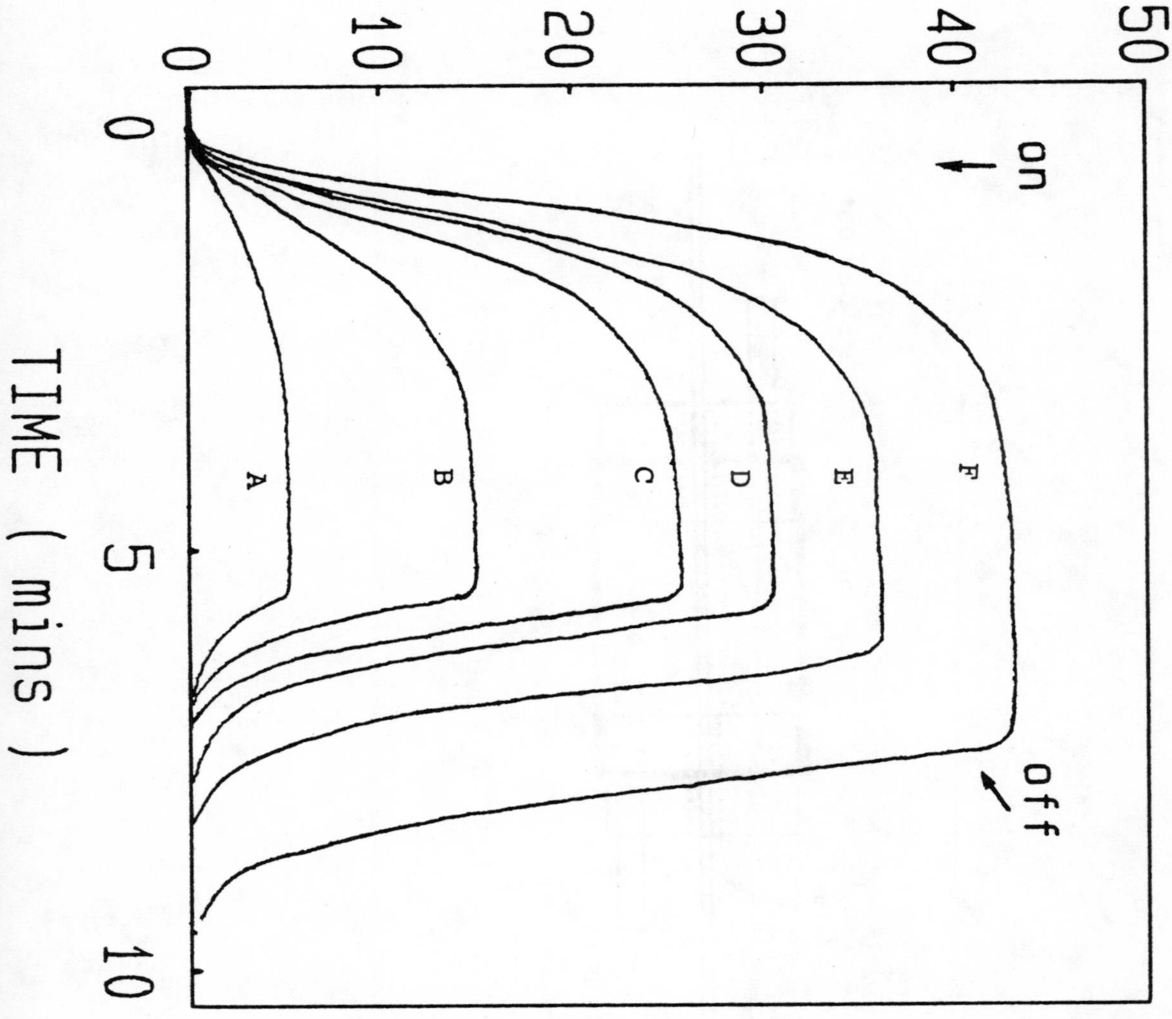
OPTICAL INTENSITY
(Arbitrary Units)
0
10
20
30
40
50
on
off
A
B
C
D
E
F
0
5
10
TIME (mins)

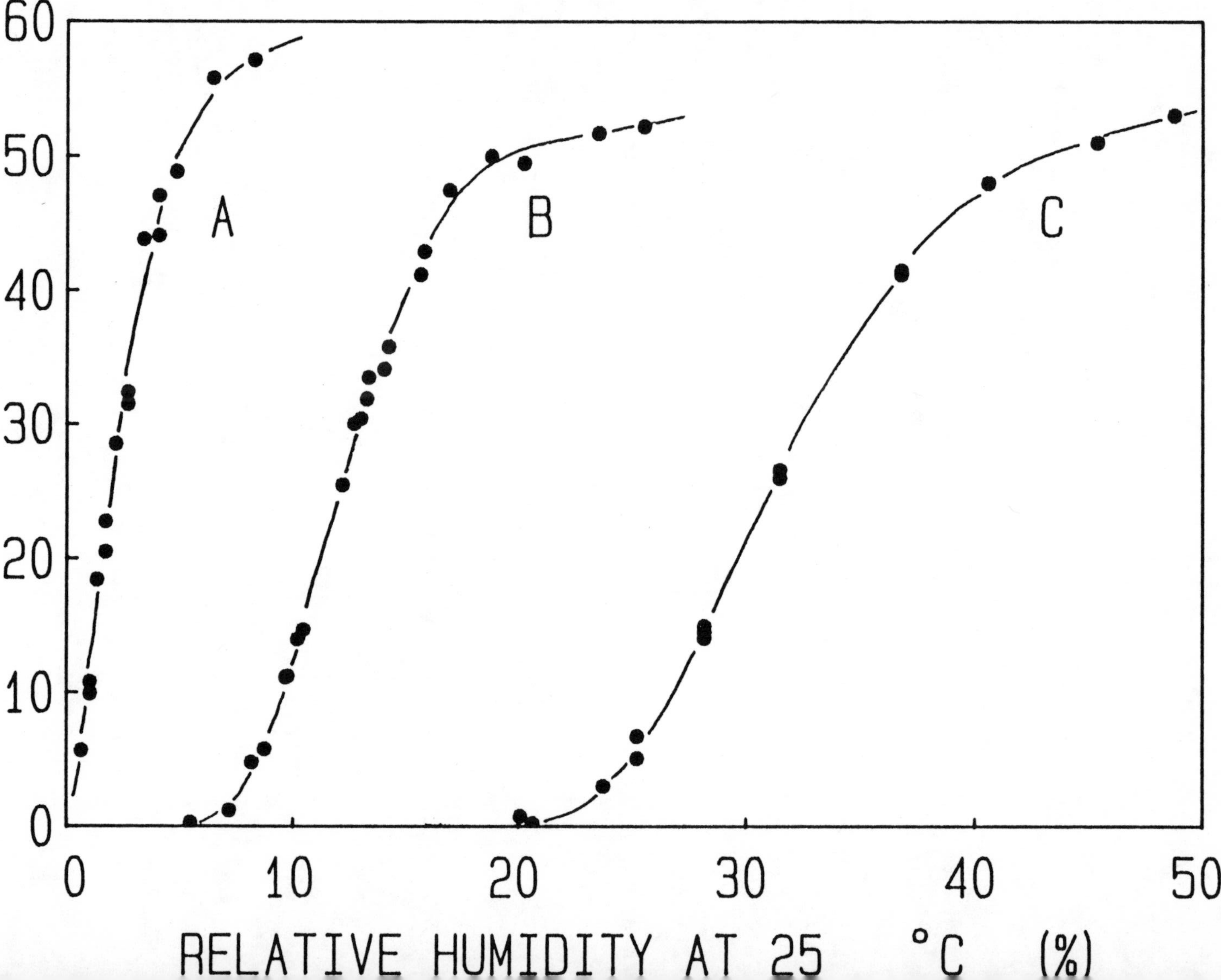
OPTICAL INTENSITY
(Arbitrary Units)
0
10
20
30
40
50
60
RELATIVE HUMIDITY AT 25 °C (%)
A
B
C

FIRDAY, JANUARY 29, 1988

MEETING ROOM 5-7-9

8:00 AM–10:00 AM

FBB1–7

ROTATION SENSORS

Ramon P. De Paula, Jet Propulsion Laboratory,
Presider

THE FIBER OPTIC GYROSCOPE : ANOTHER OPTICAL REVOLUTION IN THE INERTIA BUSINESS

Hervé C. LEFEVRE*
PHOTONETICS S.A.
52 Avenue de l'Europe 78160 MARLY LE ROI FRANCE

The fiber optic gyroscope holds a very specific position in the fiber sensor field. It is based on a purely relativistic problem of clock synchronisation which does not rely on any interaction between light and matter. It is a fascinating example of the strength of the concept of symmetry in physics : reciprocity of light propagation brings perfection to the device, despite the various defects of its components. The final optical design has converged to a great simplicity : all the potential limits, one can have thought of, have been solved through an optimized combination of system architecture and state of the art opto-electronic technologies and specific signal processing techniques.

Analysing the problem *a posteriori*, we find that, even if they have defects, guided wave components have discrete modes with different propagation constants. There are two counterpropagating primary modes which are perfectly in phase (in absence of rotation) because of reciprocity, and many waves which have been coupled into other paths with substantial optical length differences because of their discrete propagation constants. When a broadband source is used, these last parasitic waves do not produce any spurious signal because the short coherence length of the emitter drastically reduces the contrast of their interferences. Such sources add temporal filtering to the spatial and polarization filtering of the system . However some control of the components and the assembly is necessary to take full advantage of that fact. As often, the whole system used to be the best test set-up of the defects of its components, but we will detail how "white light" interferometry does provide the versatile tool which can control completely and independently the quality of the "reciprocity" of the fibergyro and its components.

Now reciprocity and the use of a broadband source improve only the quality of the bias. A fiber gyro is intrinsically an analog measuring instrument with its usual related difficulties : problems of practical scale factor accuracy and dynamic range. We will discuss the proposed solutions to reach the 10 to 100 ppm range over more than 120 dB, which is sufficient for most applications.

After more than ten years of active research throughout the world [1], the fiber optic gyroscope has actively entered its development stage. A high performance, compact and solid-state device is not a dream for system engineers any more. The nineties will see the fibergyro as a strong competitor for well-established technologies in various applications of navigation, guidance and control. It will confirm and extend the revolution brought by optics into the inertia business with active laser gyros.

* Formerly with THOMSON-CSF, LCR, 91401 ORSAY, FRANCE

REFERENCE

[1] " 10th Anniversary Conference on Fiber Optic gyros". Edited by E. Udd, SPIE Proceedings, Vol 719, Cambridge, 1986.

INTERFEROMETRIC FIBER OPTIC GRYOSCOPE USING A NOVEL 3X3 INTEGRATED OPTIC POLARIZER/SPLITTER

William J. Minford, Ramon DePaula*, Gail A. Bogert

AT&T Bell Laboratories, 555 Union Blvd., Allentown, PA 18103, (215) 439-5340

*Jet Propulsion Lab, 4800 Oak Grove Dr., Pasadena, CA 91109, (818) 354-4455

ABSTRACT

A Ti:$LiNbO_3$ 3X3 directional coupler, acting as both polarizer and 3db splitter, with a phase modulator has been incorporated in an interferometric fiber optic gyroscope. A random walk performance of $1 \times 10^{-2}\ ^0/\sqrt{h}$ has been demonstrated.

Introduction

Integrated optics brings to fiber optic gyroscopes specific advantages including a low-drive-voltage large-bandwidth phase modulator, potential low cost when the various optical functions are implemented on a single mass-produced circuit, and polarization perservation between devices on the same substrate.[1] Lefevre et al[2] have successfully used a Ti:$LiNbO_3$ multifunction circuit in a "Y-tap" configuration. The "Y-tap" configuration, however, presents a serious problem when the front-end splitter is integrated into a single integrated optic circuit; power radiating from one Y-tap can recouple into the second Y-tap.

We present the results of incorporating a 3X3 directional coupler device functioning as both the polarizer and 3dB splitter in an interferometric fiber optic gyroscope (IFOG). This splitter configuration has the potential for further integration of the front-end splitter onto the integrated optic substrate. The polarizer/splitter[3] is a 3X3 directional coupler designed with the TE polarization in the bar-state (two coupling lengths) and the TM polarization in the cross-state (one coupling length). The TM light entering the center waveguide splits into the outer guides which are connected to the gyro coil. The TE light continues out the center waveguide and is dumped.

Gyro Configuration

A schematic of the assembled IFOG is shown in figure 1. A superluminescent diode with a center wavelength of 1300nm and a bandwidth of 20nm was used as the source. The laser was thermoelectrically cooled to stablize the temperature and power. 103μW were coupled into the single-mode fiber pigtail.

The source and InGaAs PIN FET detector were attached to a 3dB polished fiber coupler made from the same polarization maintaining fiber[4] used in the sensing coil. The coupler had polarization cross-coupling less than -18dB for all ports and 0.9dB insertion loss. One output of the coupler was attached to the integrated optic substrate and the other to a power monitoring detector (not shown).

The integrated-optic circuit, a 3X3 directional coupler with a phase modulator electrode on one of the outputs, was fabricated in Z-cut y-propagating lithium niobate using Ti:indiffused waveguides for single-mode operation at λ=1300nm. A 200angstrom CVD SiO_2 buffer layer was deposited to avoid TM loading of the Al electrodes. The device, as used passively in the gyro, has a 20dB polarization extinction ratio and a 1.01 TM splitting ratio.

Non-optimum coupling of the TM mode in the 3X3 polarizer/splitter contributed 3.2dB of the 9.4dB total loss per pass of the integrated optic circuit. This TM light was guided with the TE mode in the center output waveguide. The remaining 6.2dB of insertion loss was due to fiber coupling. In order to reduce the 4% Fresnel reflections, the fiber/lithium niobate interfaces were angled. This technique[2] effectively decouples the reflections from the guided wave structure. With the 10° angling of the lithium niobate used (and therefore, the 15° angling of the fiber to account for refraction), the reflections were reduced to better than -70dB. The fiber coupling loss, 1.1dB per interface with normal end-faces, increased to 3.1 dB per interface when angled.

The gyro sensing coil, .16m in diameter, used 884 m of polarization-maintaining fiber[4], quadrapole wound. The rectangular perform-deformed high birefringence fiber coil had a polarization extinction ratio of 8.6dB (down from 19dB in the loosely wound condition) and a loss of 1.0dB.

Gyro Performance

The assembled IFOG is shown in figure 2. During the open loop testing without phase modulation and with no rotation, the returning power to the detector was 0.18μW. The 27.6dB system loss is broken down as follows: 7.8dB for the round trip through the fiber coupler, 1dB for the fiber coil and 18.8dB for the two passes through the integrated optic circuit.

The earth rate detection using the gyro is shown in figure 3. The coil was held vertically at Allentown, PA, (latitude 40°38'); therefore, the maximum earth rate is ±11.4°/h. The phase modulator was square-wave driven with ±4.5 volts to give a ± $\pi/2$ phase shift at the coil eigen frequency of 115KHz. The gyro exhibited a phase noise of $0.63(°/h)/\sqrt{Hz}$ with a lock-in amplifier time constant of 1.0 sec and 12dB/oct roll-off. This corresponds to a random walk of $0.011°/\sqrt{h}$, within a factor of three of the shot noise limit, $.0042°/\sqrt{h}$.

Conclusions

An interferometric fiber optic gyroscope was constructed using a unique integrated optic 3X3 directional coupler polarizer/splitter with a phase modulator. The gyro exhibited a phase noise of $.63(°/h)/\sqrt{Hz}$. Further improvements are expected with reduction of the integrated optic circuit loss and/or increasing the power of the source.

Acknowledgements

We wish to gratefully acknowledge those at AT&T Bell Labs who fabricated the integrated optic chip and helped in the final assembly of the gyro: Joseph Colarusso, Frederick Zwickl, Yang Chen, Kamran Bahadori, Frank Sandy and Detlef Gluszynski.

We are also grateful for the support of John McLaughlin of the Jet Propulsion Lab and Frank Stone and Robert Berry of AT&T Bell Labs.

We wish to thank Jay Simpson, AT&T-BL, for supplying the polarization maintaining fiber, Armando Martinez for wind in the coil, and C. C. Wang and William Burns of the Naval Research Lab for the superluminescent diode used in the initial evaluation.

The research was supported by the National Aeronautics and Space Administration through a contract with the Jet Propulsions Lab.

REFERENCES

[1] W. J. Minford, G. A. Bogert, and F. T. Stone, "Ti:$LiNbO_3$ Components for a Fiber Gyroscope", Proc. SPIE, vol 719, pp 146-149, 1986.

[2] H. C. Lefevre, S. Vatoux, M. Papuchon, and C. Puech, "Integrated Optics: A Practical Solution for the Fiber-Optic Gyroscope" Proc. SPIE, vol. 719 pp 101-112, 1986.

[3] G. A. Bogert, "Ti:$LiNbO_3$ Three-Waveguide Polarization Splitter". Elec. Lett., vol. 23, no. 1, pp 37-38, 1987.

[4] R. H. Stolen, W. Pleibel, and J. R. Simpson, "High-Birefringence Optical Fibers by Preform Deformation". J. Lightwave Technology, vol. LT-2, no. 5, pp 639-641, 1984.

FIGURE CAPTIONS

Figure 1 Schematic of the interferometric fiber optic gyroscope.

Figure 2 Photograph of the packaged gyroscope.

Figure 3 Earth rate response of gyroscope held vertically (Ω max = 11.4^0/h).

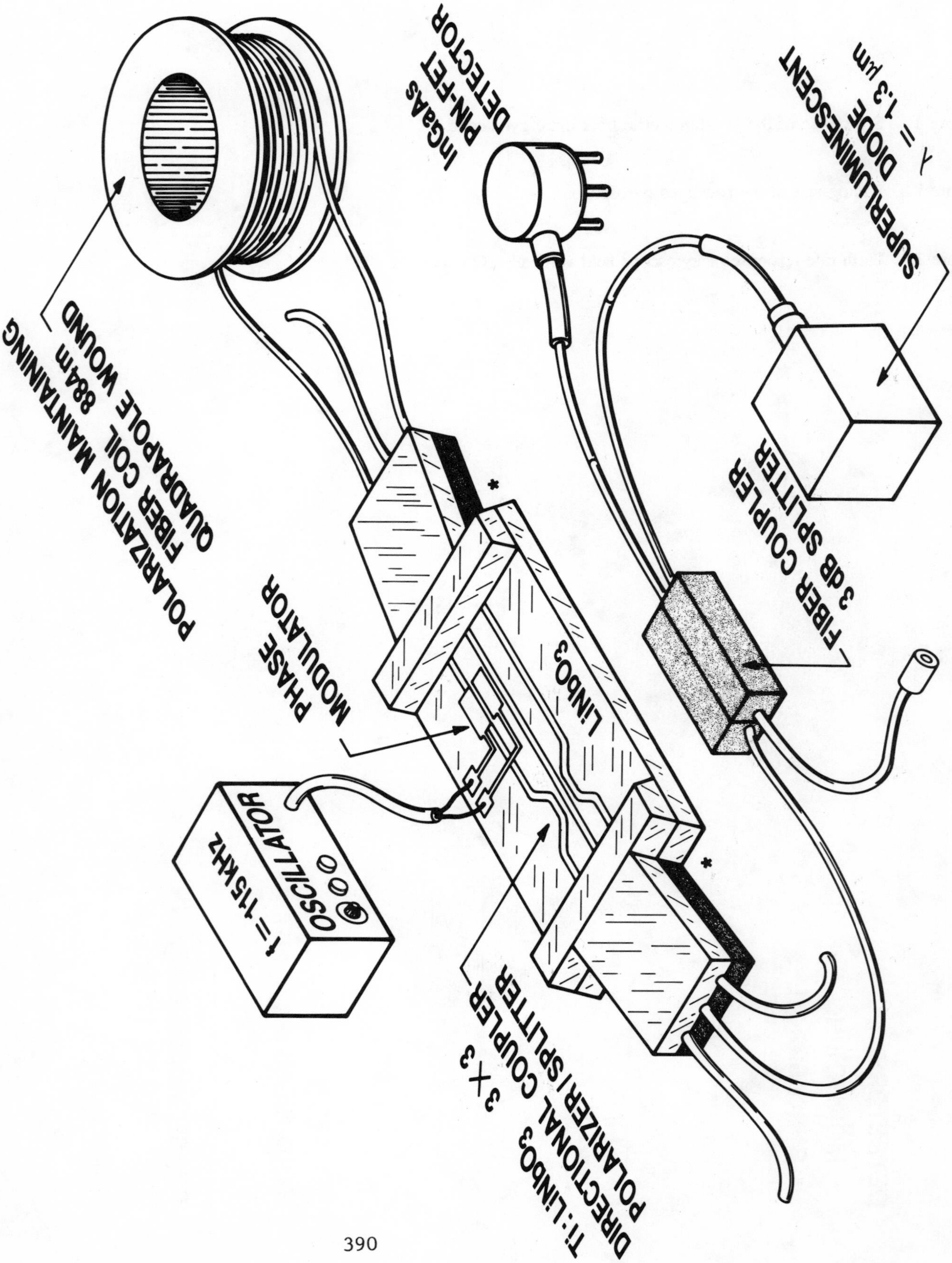

POLARIZATION MAINTAINING
FIBER COIL 884m
QUADRAPOLE WOUND

InGaAs
PIN-FET
DETECTOR

SUPERLUMINESCENT
DIODE
$\lambda = 1.3\ \mu m$

FIBER COUPLER
3 dB SPLITTER

PHASE
MODULATOR

$LiNbO_3$

OSCILLATOR
f = 115kHz

Ti:$LiNbO_3$
DIRECTIONAL 3×3
COUPLER
POLARIZER/SPLITTER

* $LiNbO_3$ / FIBER INTERFACES ANGLED TO
REDUCE REFLECTIONS TO < -50 dB LEVEL

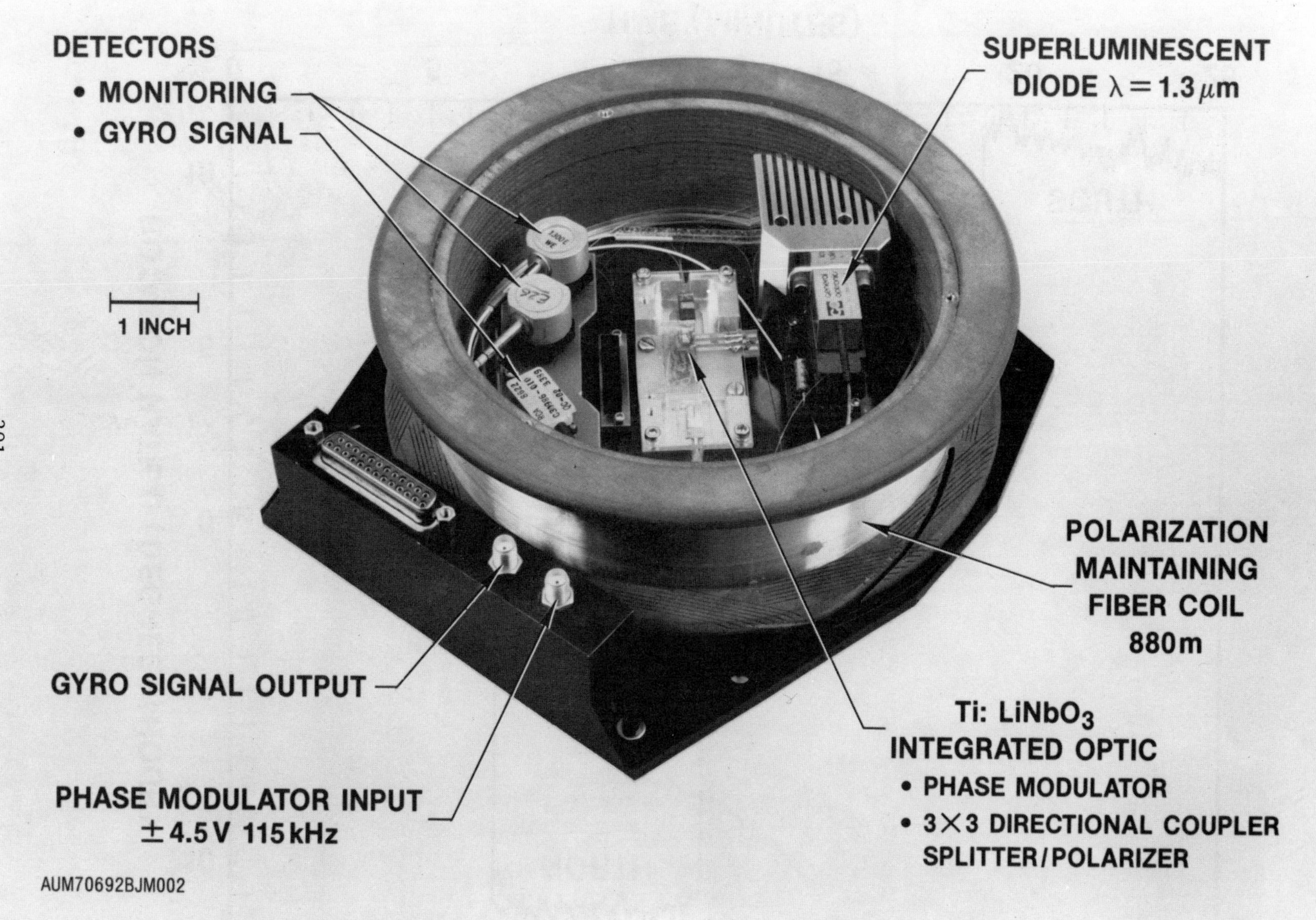

DETECTORS
• MONITORING
• GYRO SIGNAL
1 INCH
GYRO SIGNAL OUTPUT
PHASE MODULATOR INPUT
±4.5 V 115 kHz
AUM70692BJM002
SUPERLUMINESCENT
DIODE λ = 1.3 μm
POLARIZATION
MAINTAINING
FIBER COIL
880 m
Ti: LiNbO3
INTEGRATED OPTIC
• PHASE MODULATOR
• 3×3 DIRECTIONAL COUPLER
SPLITTER/POLARIZER

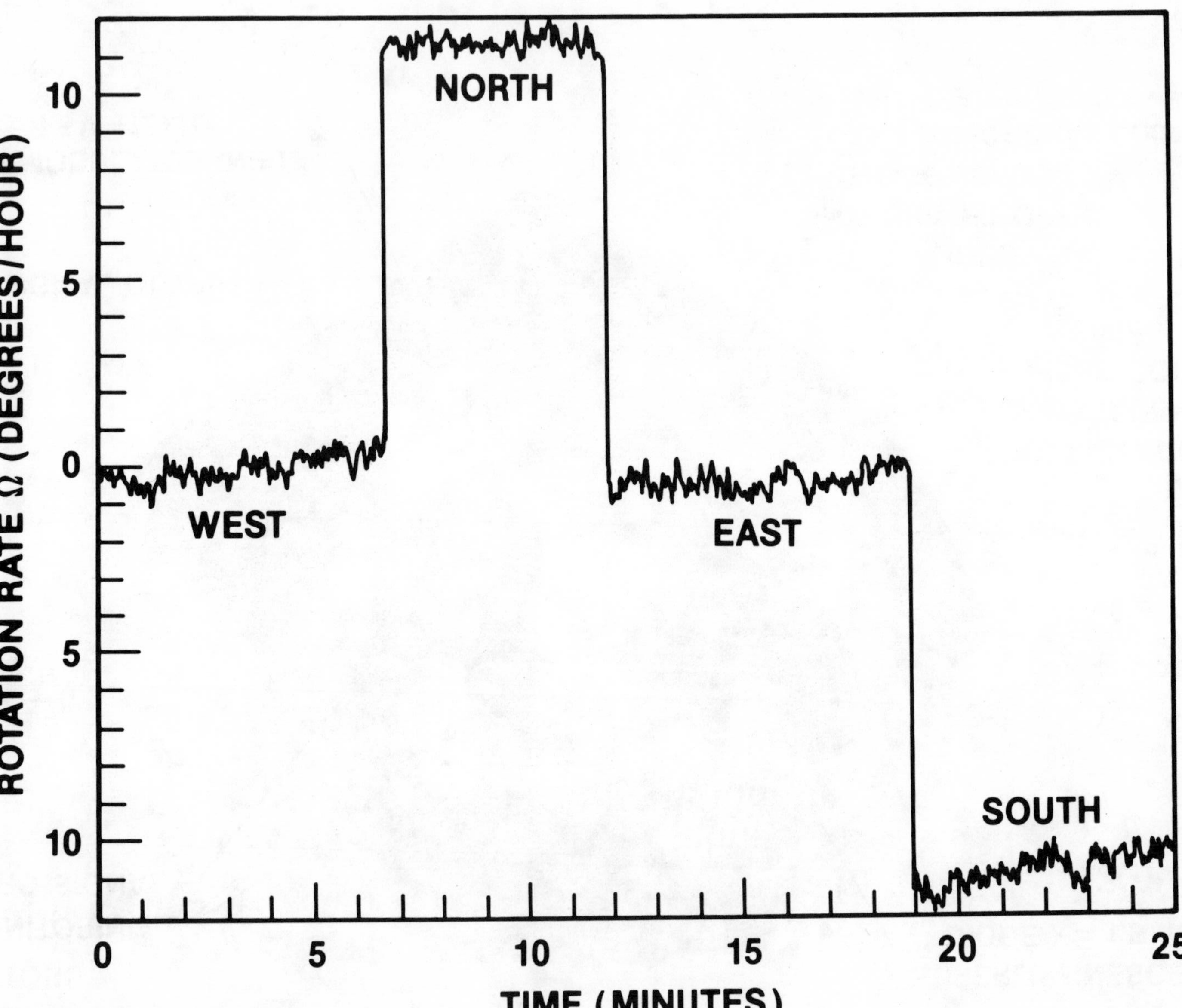

AUM70692BJM001

Scale Factor Accuracy and Stability in an Open Loop Fiber Optic Gyroscope

R. P. Moeller, W. K. Burns and N. J. Frigo
Naval Research Laboratory
Washington, DC 20375-5000
202-767-3298

SUMMARY

1. Introduction

To date, the most sensitive and accurate passive optical gyroscopes have been phase reading interferometers with optical outputs of the form $I = I_0 (1 + \cos \phi_{tot})$, where ϕ_{tot} is the non-reciprocal phase shift due to all sources. Ideally, the phase shift has only two components: $2\phi_s$, the Sagnac phase shift (proportional to the rotation rate), and ϕ_{dither}, which is introduced by a transducer to facilitate the measurement. In the gyroscope we report here, ϕ_{dither} is applied to a "minimum configuration" gyroscope and creates an ac output of the form

$$I = I_o \cos\phi_{tot} = I_o (-\sin 2\phi_s (J_1(x) \cos \omega_m t + \ldots)) + I_o (\cos 2\phi_s (J_2(x) \cos 2\omega_m t + \ldots)) \quad (1)$$

where ω_m is the frequency of the piezoelectric transducer, and x is the amplitude of the non-reciprocal phase shift caused by it. If the output were described exactly by Eq.1, then the voltage output of the detector at the frequency ω_m would be sinusoidal in rotation rate. To a high degree of approximation this is the case, but we discuss corrections which, when taken into account, will improve the accuracy of the subsequent scale factor determination.

II. Coherence and feedback effects

The ideal output described in Eq. 1 is not actually realized. Since the Sagnac and transducer phase shifts create unbalanced path lengths, the limited coherence length of the source makes the time average of the intensity decrease with rotation rate[1] and dither. These two effects ("dc" and "ac" coherence effects) will be described. Additionally, because of the high optical quality of the gyroscope, optical feedback takes place and influences the output. Depending on rotation rate and transducer excursion, optical feedback to the SLD will vary, influencing the optical output to the gyroscope. The differential optical output (i.e. compared to the ideal situation of no optical feedback) is proportional to $\cos(\phi_{tot})$ and thus I_o in Eq. 1 has a sinusoidal dependence on rate and dither. This mixes in Eq. 1 to produce a term proportional to $\sin 4\phi_s$ in the demodulated output.

III. Gyro design and baseline performance

The optical configuration for the all fiber gyroscope is similiar to that reported previously[2]. A piezoelectric phase shifter is employed in the usual way for biasing. The gyro used all polarization preserving fiber with fusion splices between the various components. The total gyro insertion loss, including 6 dB of unavoidable circuit loss, was 14 dB. An electrical feedback circuit was used to stabilize the source output power.

The gyro coil (R=16cm, L=1km) and optical components were placed in a large temperature controlled box and mounted on a rate table. The measured random walk coefficient was 3.2×10^{-4} deg/hr$^{1/2}$. The noise equivalent rotation rate (0.4 sec time constant) was below 0.01 deg/hr for an optical power of 1 mW leaving the SLD (λ=0.83μm). Bias stability tests were also performed on the gyro and will be reported.

IV. Data Retrieval and Open Loop Output

Our experimental test setup consisted of a rate table driven by a desktop computer. With the piezoelectric phase shifter in the loop, the expected form of the detected output after demodulation at ω_m is a voltage

$$V(\Omega) = A_1 \sin (A_2 \Omega + A_3) + A_4 \qquad (2)$$

where A_1 is an "electrical" scale factor, A_2 is an "optical" scale factor, A_3 is an optical offset (or bias), and A_4 is the electrical bias in the demodulation circuit. By rotating the gyro at a series of rates, data is obtained to which Eq. 2 may be fit by a least squares technique and the 4 coefficients of Eq. 2 can be determined. After such a "calibration", Eq. 2 can then be inverted to yield rotation rates for raw voltages.

Each of these computed rates can then be plotted against the known input rate and compared to the expected linear fit. The quality of this fit, which reflects the accuracy of the computed scale factor coefficients, is described by a computed rms error. From our discussion above, we expect coherence and feedback effects to cause deviations from the simple sinusoid of Eq. 2. To demonstrate the effects of these phenomena we take a typical run and sequentially analyze the data, successively applying additional correction factors. To judge the fit to Eq. 1, we look at the difference between the measured and expected voltages of the least squares fit to Eq. 2. These are the residuals after a fit is performed. In Fig. 1a we show the initial fit results without any correction factors. The rms error characterizing the fit is shown in the figure. In Fig 1b we show the result of a dc coherence correction, i.e., the raw data is corrected by the source coherence function. We see that the residual voltage differences clearly show the $\sin 4\phi_s$ dependence expected to arise from the feedback effect. In addition to the dc coherence correction discussed above, we applied a feedback correction by adding the term proportional to $\sin 4\phi_s$ to the measured voltages and obtained the results in Fig 1c. The residuals are is reduced to less than 1 mV and the rms error to 1.5 deg/hr. Clearly, these two corrections have improved the fit of the measured data to Eq. 2.

V. Scale Factor Stabilities

To operate as an effective navigation tool, the gyroscope's scale factors and offsets must be stable over time. We have examined some of the environmental influences on the coefficients in Eq. 2.

We ran a series of test runs, with rates up to +/- 4 deg/sec, over a period of 12 days to test short term (i.e. tens of days) scale factor stability. For each run, (there were several each day) the scale factor coefficients of Eq. 2 were computed as described previously. The results for A_1 and A_2 are plotted in Figs. 2 and 3. A_1 was found to depend on the temperature of the lockin amplifier and A_2 on the SLD drive current. These dependencies are taken into account by fitting a line through the plot of the coefficient vs. temperature or current. The rms deviation (1σ) of the data from these fit lines are 32 ppm for A_1 and 16 ppm for A_2, showing good stability over the 12 day period. The biases A_3 and A_4 drifted within 2.5 deg/hr for A_3 and 1.2 deg/hr for A_4, although A_4 was also correlated with the lockin temperature.

VI. Conclusion

We have isolated electrical and optical contributions to the various scale factors and offsets present in an open loop fiber optic gyroscope, and have shown the effects of optical source coherence and optical feedback on the functional form of the gyro output. This work has, for the first time, demonstrated inertial quality scale factor stability in an open loop gyro, and may open up new application areas for fiber optic gyros.

REFERENCES

1. W. K. Burns and R. P. Moeller, JLT, LT-5, 1024 (1987).
2. W. K. Burns, et. al., Optics Lett. 9, 570 (1984).

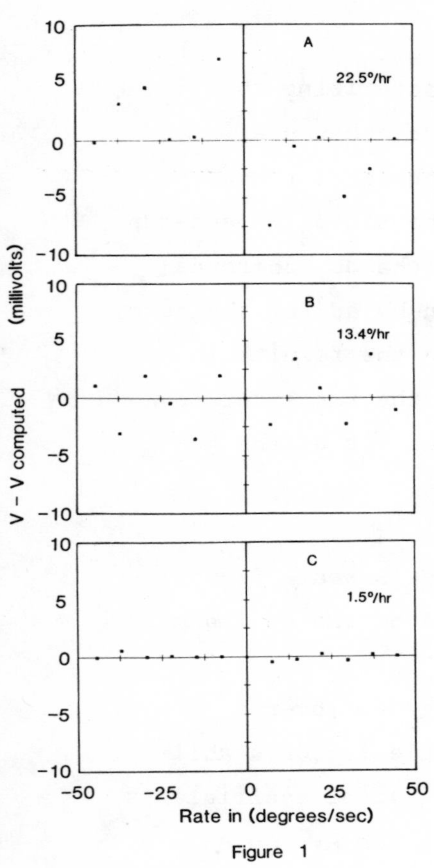

Figure 1

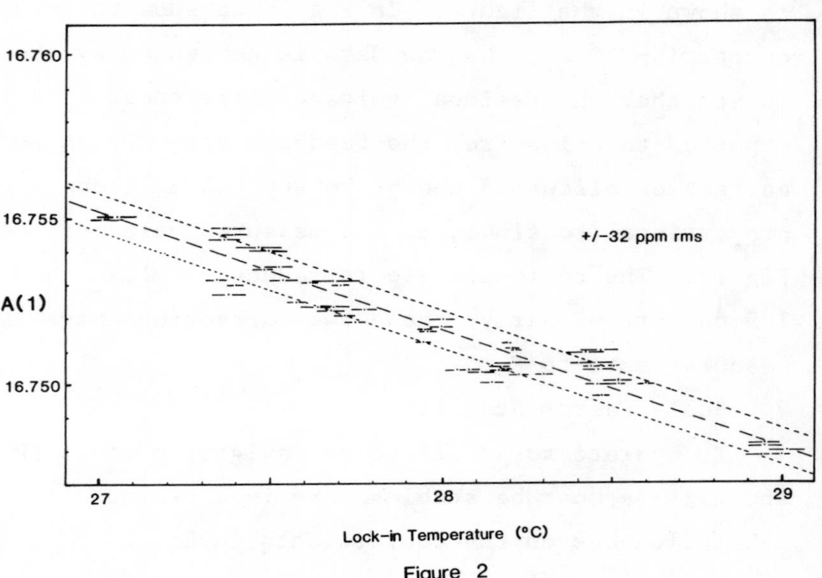

Figure 2

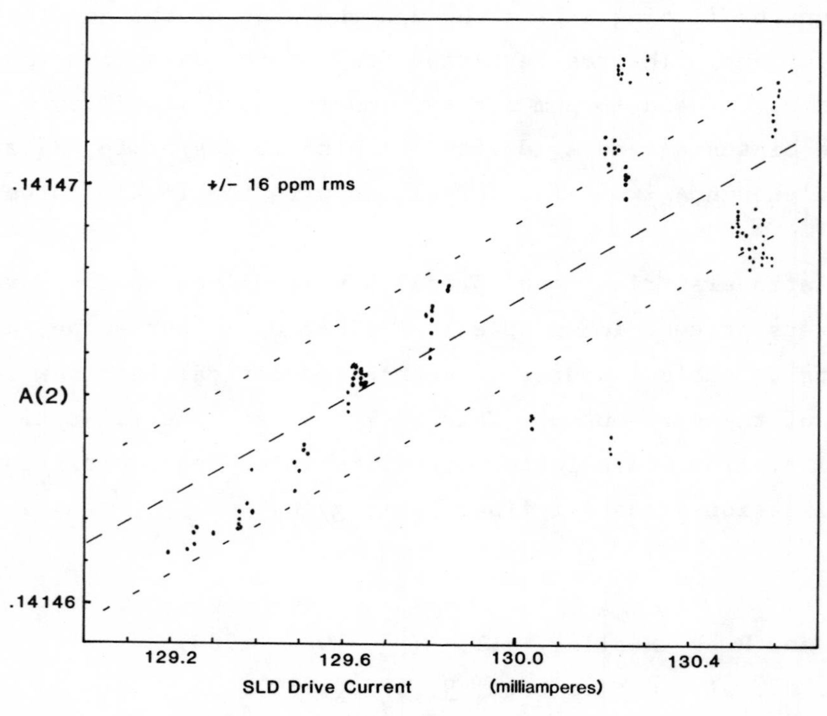

Figure 3

DRIFT REDUCTION IN THE OPTICAL HETERODYNE FIBER GYRO

Kazuo HOTATE and Shigeatsu SAMUKAWA

The University of Tokyo
Research Center for Advanced Science and Technology

4-6-1 Komaba, Meguro-Ku, Tokyo 153, JAPAN
Phone 03-481-4437
Facsimile 03-485-5135

Noboru NIWA

Chiba Institute of Technology

2-17-1 Tsudanuma, Narashino-Shi, Chiba 275, JAPAN
Phone 0474-75-2111

1. Introduction

Optical Heterodyne Fiber Gyro[1,2], which we have proposed and experimented with, has a wide dynamic range and a good linearity. We demonstrated the rotation detection with 5 digits[1,2]. However, it has a relatively large zero-point drift induced mainly by thermal fluctuation of the acoustic-wave velocity in the Acousto-Optical Modulator (AOM)[1,2].

In this paper, we propose an improved optical system and a signal processing scheme for the Optical Heterodyne Fiber Gyro to overcome the above problem. Experimental results demonstrate successfully the reduction of the zero-point drift.

2. Description of the Improved Optical Heterodyne Fiber Gyro

Figure 1 shows a conceptual diagram of the improved Optical Heterodyne Fiber Gyro for the drift reduction. We introduce the reference path in the optical system and develop a new signal processing scheme.

Two waves, divided by a grating and a beam splitter, are launched into the signal path and the reference path to propagate in the opposite derections in each path. Two waves emitted at the fiber ends are recombined by AOM. The grating and the AOM are located symmetrically with respect to the beam splitter. Detector D receives a nondiffracted part (frequency f_0, equal to the laser-oscillation frequency) of the clockwise(CW)-propagated wave and a diffracted part ($f_0 + f_1$, where f_1 is the AOM driving frequency) of the counterclockwise(CCW)-propagated wave, then outputs the intermediate frequency f_1.

To distinguish the signal lights traveled in the sensing fiber from the reference lights, the light source is modulated in a square form with the cycle of twice the round-trip time in the sensing fiber loop. When the detector output is divided into two parts using an electronic switch synchronized with the modulation, we can obtain the signal including the Sagnac phase shift and the reference not including the phase shift.

In the basic configuration[1,2], the AOM driving signal is used as a phase reference. Therefore, the thermal drift of the acoustic wave velocity in the AOM was the drift source. The improved system proposed in this paper can avoid this drift.

The output signals of the switch are not continuous wave. To convert the signals to continuous waves without changing the phase information, the signal processing circuit of Fig.2(a) is developed. Each output phase of the switch is compared with the continuous wave from the AOM driving wave by double-balanced mixer(DBM). The phase of the continuous wave is controlled by the phase shifter to minimize the DBM's output. Then the continuous output of the phase shifter has the same information as each output of the switch. By measuring the phase difference between the signals from the two phase shifters using the phase-nulling detection circuit as shown in Fig.2(b), which was developed for the basic configuration[1,2], we can obtain the rotation rate. The output of the circuit is electronic frequency proportional to the input rotation rate. Then, we can obtain so wide dynamic range[1,2].

In the improved configuration, the themal fluctuation of the acoustic-wave velocity in the AOM affects both the signal light and the reference light in the same way. Therefore, the drift can be canceled. Though this configuration does not show the reciprocity between the CW and CCW signal lights, it shows reciprocity between the signal light and the reference light.

3. Experiment

In the experimental setup, the sensing fiber is a 1,940m long single-mode fiber wound on a 15.9cm radius drum. The light source is a single-mode laser diode with 780nm wavelength. Its modulation frequency is 52.7KHz. The detector is APD.

Figure 3 shows the zero-point drift in 5 minutes after the AOM driving signal is jointed to the AOM from the dummy load. Figure 3(a) shows the phase change of the signal lights compared with the AOM driving wave. This output corresponds to that of the basic configuration[1,2]. Figure 3(b) shows the phase change of the reference lights. Figure 3(c) shows the phase difference between Figs.3(a) and 3(b), which is the output of the improved configulation. In Fig.3(a), that is, in the basic configuration, the zero-point drift by the thermal fluctuation of the acoustic wave is about 0.2deg/s. However, in the improved configuration such large zero-point drift is sufficiently reduced as shown in Fig.3(c).

This experiment demonstrates that the improved Optical

Heterodyne Fiber Gyro can reduce the drift effectively. The short time resolution in this experiment does not limited by the system configuration itself, but limited by the coherent noise because of the use of the single-mode LD.

Figure 4 shows the confirmation that the setup has the gyro function. Only the fiber coil is rotated by hand in a short time. The dynamic range and the lineaity of the Optical Heterodyne Fiber Gyro has already been demonstrated[1,2].

4. Conclusion

We have developed the improved Optical Heterodyne Fiber Gyro. The reference path has been introcuced in the optical system, and the signal processing scheme has been adopted. We have successfully reduced the zero-point drift in the experimental setup.

References

[1] K. Hotate, N. Okuma, M. Higashiguchi and N. Niwa, "Rotation detection by Optical Heterodyne Fiber Gyro with frequency output," Optics Lett., 7(1982),331.

[2] -----,"Optical Heterodyne Fiber Gyro with frequency output," Symp. Gyro Tech., Stuttgart(Sept. 15-16, 1982), 4.1.

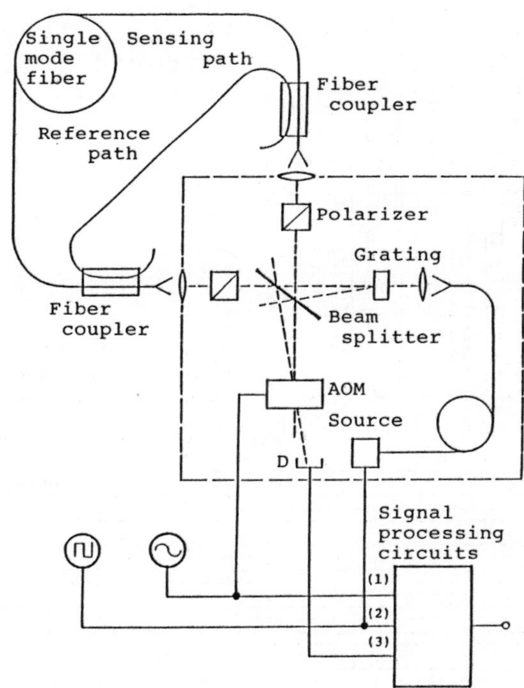

Fig.1 Conceptual diagram of the improved Optical Heterodyne Fiber Gyro with the drift reduction.

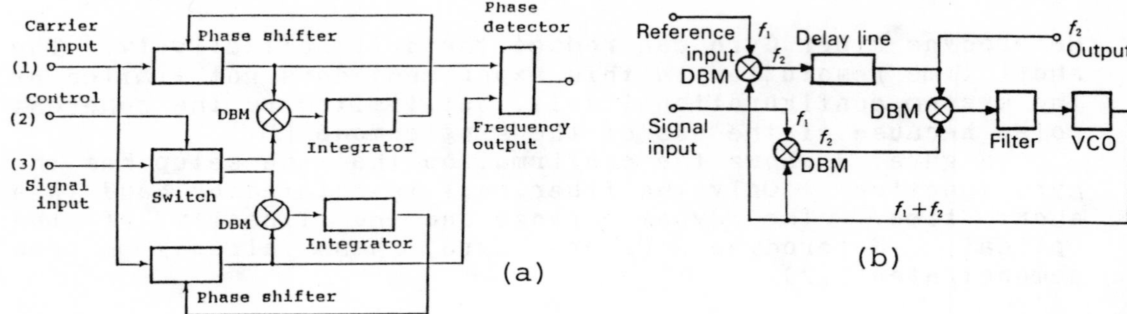

Fig.2 Signal processing circuits. (a)Circuit for making the continuous waves. (b)Circuit for phase-nulling detection with frequency output.

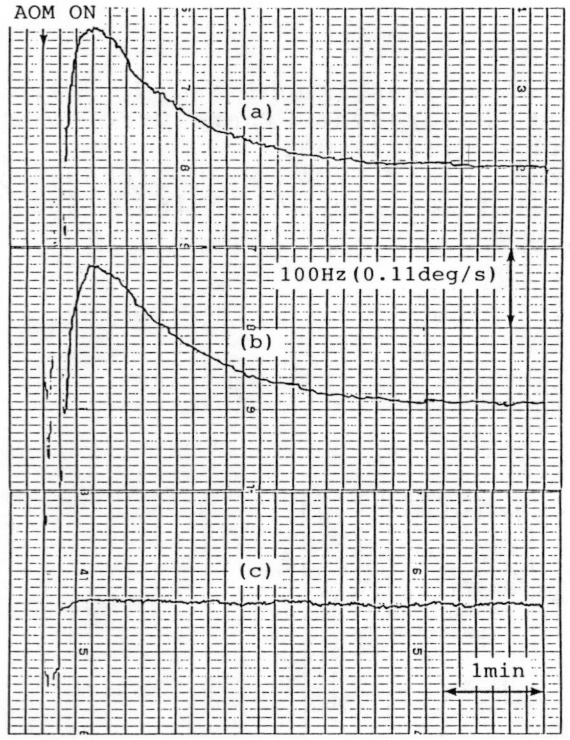

Fig.3 Zero-point drift by thermal variation in the AOM just after the switch-on of the AOM. (a)Phase change of the signal light, (b)phase change of the reference light, (c)output of the improved Optical Heterodyne Fiber Gyro.

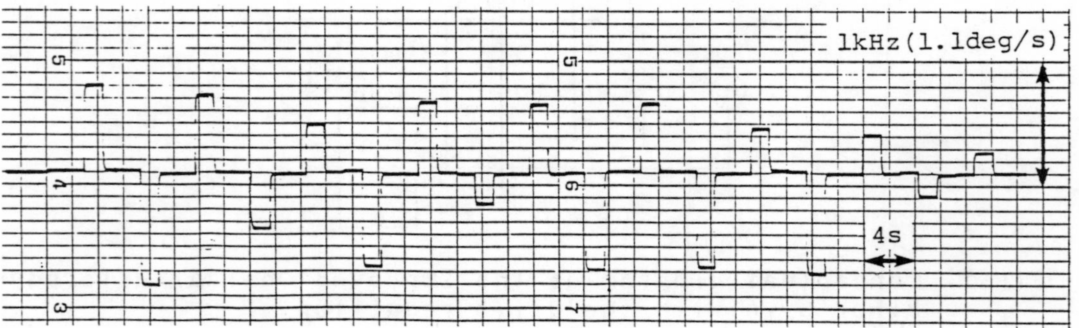

Fig.4 Rotation detection by the experimental setup of the improved Optical Heterodyne Fiber Gyro.

EFFECT OF REFLECTIONS ON THE DRIFT CHARACTERISTIC OF A FIBER-OPTIC PASSIVE RING-RESONATOR GYRO

MASANOBU TAKAHASHI, SHUICHI TAI, KAZUO KYUMA, AND KOICHI HAMANAKA

CENTRAL RESEARCH LABORATORY
MITSUBISHI ELECTRIC CORPORATION
8-1-1 TSUKAGUCHI-HONMACHI,
AMAGASAKI, HYOGO, 661 JAPAN
PHONE: 06-491-8021

1. Introduction

A fiber-optic passive ring-resonator gyro (FOPRG)[1)] has an advantage of reducing a fiber length over a conventional fiber-optic interferometric gyro. However, the FOPRG has several peculiar problems to be solved in order to achieve the high sensitive detection of the rotation rate. Among them, reducing the reflection is very important. For example, it was reported that the backscattered light generated in the fiber ring splits the resonant peak into two, resulting in the degradation of the FOPRG linearity.[2)]

On the other hand, in this paper, it is pointed out for the first time that the reflected light occurred outside the fiber-ring yields a serious degradation on the drift characteristic. Such an effect of reflection is shown to be removed by an external optical phase modulator. This point is particularly discussed in detail. The second new idea incorporated is the use of an external-cavity laser diode as a compact high-coherence light source. As a result of these efforts, the detection sensitivity as low as $3X10^{-5}$rad/sec is obtained by the FOPRG with 10cm in a fiber-ring diameter.

2. Effects of the reflected lights

A schematic diagram of the constructed FOPRG is shown in Fig.1. The two counterpropagating lights along the fiber-ring are respectively used as a sensing light and a reference light whose output is fedback to the PZT(1) so that the fluctuations in the resonant peak caused by the external disturbances are compensated. As for Fig.1, the reflected lights at the fiber edges are returned to the fiber ring and interfered with the sensing and reference lights. The resonant characteristic of the fiber-ring resonator is varied by the intensity and phase fluctuations of the reflected lights. Fig.2 shows the calculated resonant characteristic of the fiber-ring resonator in which the reflectivity R at the fiber edges are assumed to be 4% due to the Fresnel reflection. θ is the phase difference between the signal light and the reflected light. As θ changes very sensitively owing to the unavoidable little external disturbances, the resonant characteristic undergoes a large variation, which causes a serious drift of the FOPRG output. Fig.3 shows the observed variations in the resonant characteristic for the reflectivity of 4%. As compared with Fig.2, we can see that a close agreement between the observed and calculated values is obtained. Fig.4 shows the calculated maximum drift width Ω_d as a function of the phase modulation amplitude ϕ_m by the PZT(1), where Ω_d is normalized by R (R<<1) and ϕ_m by δ , the FWHM value of the resonant peak. It can be shown theoretically that the highest detection sensitivity is obtained at $2\phi_m/\delta$ =0.8. Even at this condition, Fig.4 indicates that the reflectivity of about 10^{-14} is required to obtain the sensitivity of 10^{-6}rad/sec. However, this small value of reflectivity is difficult to achieve from the practical point of view. One method to solve this problem is the use of the optical phase modulation technique[3)] together with the efforts of reducing the reflected light. By modulating the phase of the light at a proper location, the reflected lights are shifted in frequency from the sensing and the reference lights, and then the noise components arisen from the reflections can be decoupled electrically. It should be noted that the phase modulator is located before the reflection point.

3. Experiments

As shown in Fig.1, the phase modulator for this purpose was constructed of the mirror mounted on PZT(2). Another new point of this paper is that an external-cavity laser diode[4)] is used as a high coherence light source, as described previously. The external-cavity laser has an advantage of more compact light source than the He-Ne laser. The cavity length of the constructed external-cavity laser was about 6cm and its spectral linewidth was 100kHz. The fiber-ring resonator was constructed of a polarization maintaining fiber and a low-loss fiber directional coupler. Its fiber length, diameter and finesse were 4m, 10cm and 30, respectively. The two end faces of fiber edges were polished at an oblique angle of 7° to reduce the reflections. The maximum drift width measured was about 10kHz for the FOPRG without the phase modulator (PZT(2)). From this result, the reflectivity at the fiber edges was estimated about 10^{-4}. Fig.5 shows the drift characteristic when the optical phase modulation is applied. The time constant of the lock-in-amplifier(2) (LIA(2)) was 30sec. The RMS frequency noise of about 2Hz, which corresponds to the detection sensitivity of about $3X10^{-5}$rad/sec for the FOPRG with the fiber-ring diameter of 10cm, was obtained. The residual drift is perhaps caused by the deviation from the optimum phase modulation amplitude of PZT(2). The effect of multi-reflections between the two fiber edges, which can not be removed perfectly by the phase modulation technique, might be the origin of the residual drift.

4. Conclusion

We have shown theoretically and experimentally that the reflections in the light path outside the fiber ring results in a serious effects on the FOPRG performance. The phase modulation technique was found to be useful to reduce the effect of such reflections. It was pointed out that the optical phase modulator should be located outside the fiber. Furthermore, an external-cavity laser diode was first used in the FOPRG. The detection sensitivity as low as $3X10^{-5}$rad/sec (2Hz) was achieved with an integration time of 30sec.

Acknowledgments

The authors would like to thank Y.Mitsuhashi of Electrotechnical laboratory and T.Nakayama of our laboratory for their continuous encouragement.

Reference

(1) R. E. Meyer, S. Ezekiel, D. W. Stowe and V. J. Tekippe: "Passive fiber-optic ring resonator for rotation sensing", Opt. Lett. 8,pp.644-646(1983).

(2) K. Iwatsuki, K. Hotate and M. Higashiguchi:"Backscattering in an optical passive ring-resonator gyro : experiment", Appl. Opt. 25,pp.4448-4451(1986).

(3) G. A. Sanders, M. G. Prentiss and S. Ezekiel:"Passive ring resonator method for sensitive inertial rotation measurements in geophysics and relativity",Opt. Lett. 6, pp.569-571(1981).

(4) M. W. Fleming and A. Mooradian:"Spectral characteristics of external-cavity controlled semiconductor lasers", IEEE J. Quantum Electron. QE-17,pp.44-59(1981).

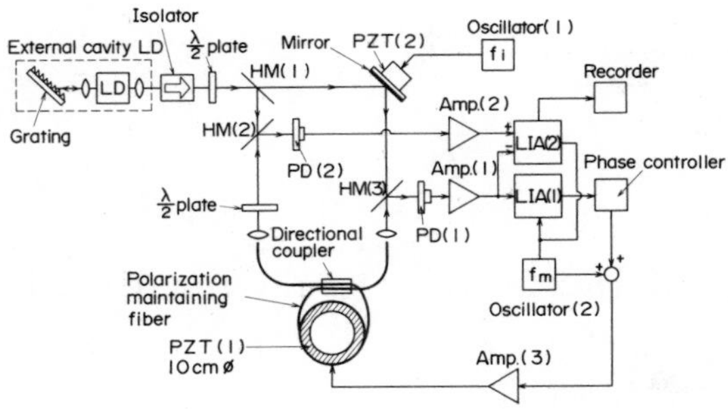

Fig.1 Schematic diagram of the constructed FOPRG

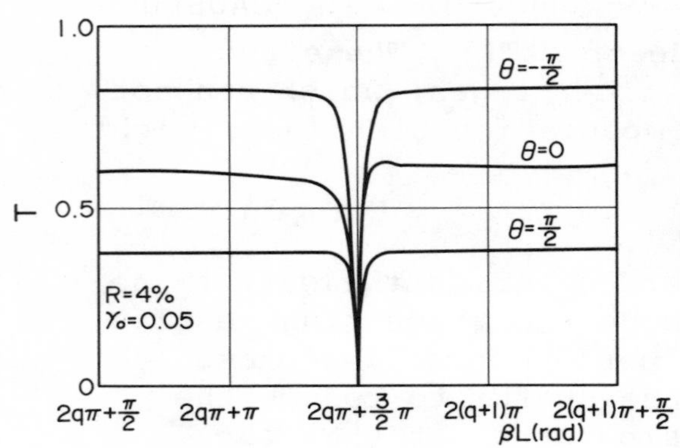

Fig.2
Calculated variation of the resonant characteristic of the fiber-ring resonator

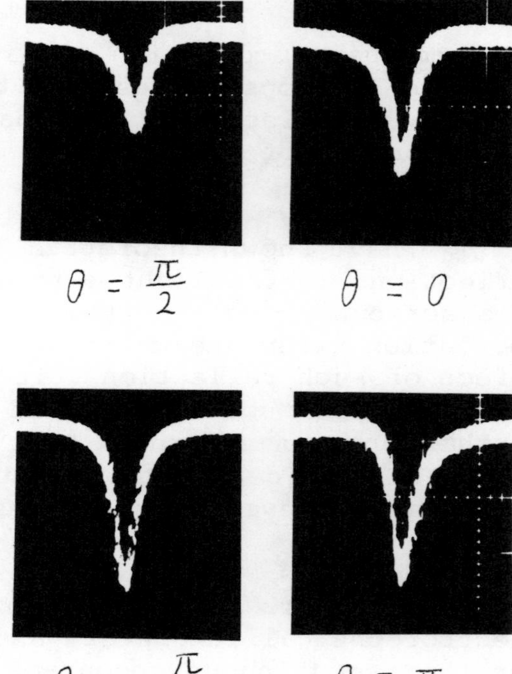

Fig.3
Observed variation of the resonant characteristic of the fiber-ring resonator (R=4%)

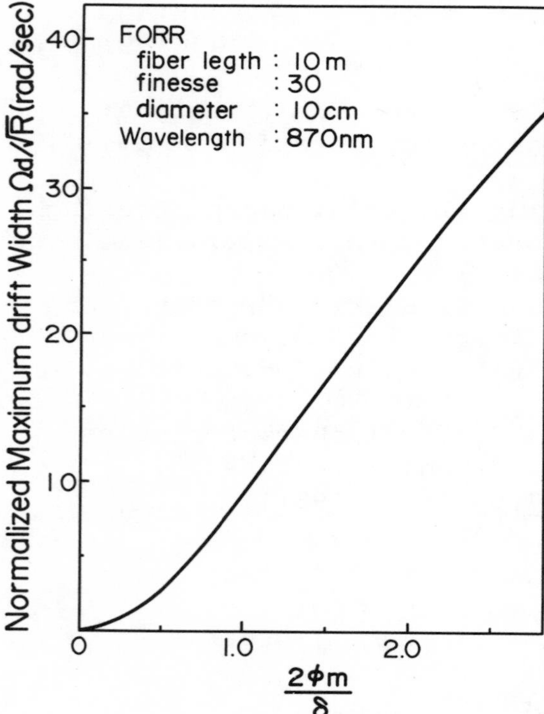

Fig.4
Calculated normalized maximum drift width as a function of the normalized optical phase modulation amplitude

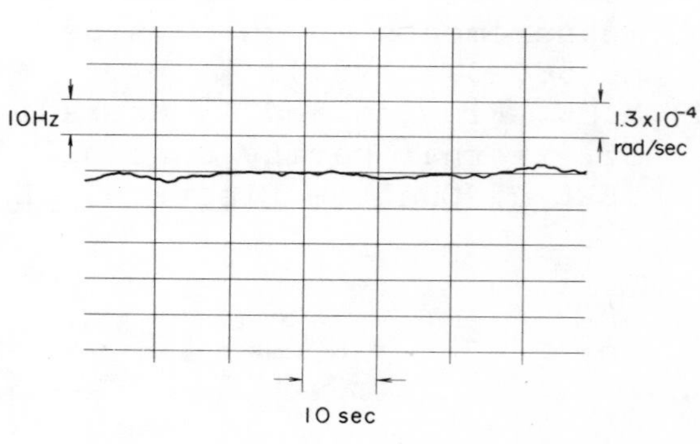

Fig.5
Measured drift characteristic

DRIFT OF AN OPTICAL PASSIVE RING-RESONATOR GYRO CAUSED BY THE FARADAY EFFECT

K. HOTATE AND M. MURAKAMI

THE UNIVERSITY OF TOKYO
RESEARCH CENTER FOR ADVANCED SCIENCE AND TECHNOLOGY

4-6-1 KOMABA, MEGURO-KU, TOKYO 153, JAPAN
PHONE 03-481-4437
FACSIMILE 03-485-5135

1. Introduction

Noise sources in the optical passive ring-resonator gyro(OPRG) must be completely estimated and eliminated for practical applications. Some of those noise sources have been investigated and their reducing procedures have been reported [1 - 4]. The influence of the magnetic field is one of the noises, which must be clarified.

In this paper, we present the results of the theoretical analysis on the drift caused by the Faraday effect which is attributed to the earth's magnetic field. It is clarified that the drift can be suppressed by using a polarization maintaining optical fiber and by reducing a special twist component in the sensing fiber loop.

2. Faraday Effect in OPRG

Figure 1 shows the basic configuratian of an optical passive ring-resonator gyro. We must use two counterpropagating lights to compensate the thermal and/or the mechanical fluctuations. When this OPRG is located in a magnetic field ,its fiber loop acts as a Faraday device and polarization plane of the linearly polarized light in the fiber rotates. Generally, the phase difference appers between the CW and CCW light travelling in the fiber loop by the nonreciprocity of the Faraday effect when retarders are situated at asymmetric points in the fiber loop [5]. In short,the presence of both the magnetic field and the retarders in the fiber is the condition for the apperance of the drift caused by the Faraday effect. In a practical fiber ,retarders due to small birefringence,twist,and so on are located at random along it. They generate the output drift in cooperation with the Faraday effect.

3. Analysis and numerical evaluation

In this section we consider the twist of the fiber as an actual and inevitable retarder to analyze the drift when using a polalization maintaining fiber.

In general, two eigenstate of polarization(ESOP) exist in the ring resonator, and one ESOP is usually orthogonal to the other [3]. As a result, each ESOP generates the resonance independently. The resonance point of each ESOP is concerned with the eigenvalue of transfer matrix of the whole fiber loop including the coupler. In the following, we calculate it when a fiber loop with random twist is located in a magnetic field.

Under the assumption that $\Delta\beta \gg \varphi$, ζ_0 denoting the fiber birefringence, the twist per unit length, and the Faraday rotaion per unit length as $\Delta\beta$, φ , and ζ_0, respectively, the eigenvalues for the CW light are calculated as

$$\lambda^{cw}_{1,2} = \exp\left[-j\beta_{av}L \pm \int_0^L \eta^{cw}(z)\,dz\right], \quad (1)$$

where

$$\eta^{cw}(z) = \sqrt{\left(\frac{\Delta\beta}{2}\right)^2 + \left\{\varphi(z) + \zeta_0 \sin\frac{2\pi}{L}z\right\}^2}, \quad (2)$$

L is the fiber loop length, β_{av} is the average propagation constant of the fast and slow mode, and z is the coordinate along the fiber length.
For the CCW light, we obtain the equation:

$$\lambda^{ccw}_{1,2} = \exp\left[-j\beta_{av}L \pm \int_0^L \eta^{ccw}(z)\,dz\right], \quad (3)$$

$$\eta^{ccw}(z) = \sqrt{\left(\frac{\Delta\beta}{2}\right)^2 + \left\{\varphi(z) - \zeta_0 \sin\frac{2\pi}{L}z\right\}^2}. \quad (4)$$

The double sign in Eq.(1) and (3) corresponds to the existence of two eigenstate of polarization in the ring resonator.Futhermore, a difference in the sign between Eq.(2) and (4) indicates the nonreciprocity of the Faraday effect. The phase of the eigenvalue is equal to the round-trip phase of the light in the resonator. As a result the difference in the round-trip phase between the CW and CCW light, which corresponds to the resonance-frequency difference, is given by

$$\Delta\psi = \frac{4\zeta_0}{\Delta\beta}\sqrt{2AW}\cos\theta, \quad (5)$$

where

$$W = \left|\int_0^{2m\pi} \varphi(\theta)\exp(-j\theta)\,d\theta\right|^2 \Big/ 2m\pi, \quad (6)$$

A, m, and θ is the area surrounded by the fiber loop ,turn number of the fiber loop, and the direction of the magnetic field, respectively. In eq.(6) W means the fiber twist component whose period is just equal to one turn of the fiber loop. The resonance frequency diffrence corresponds to Eq.(1)-(4) is illustratedin Fig.2. When only the twist is applied in the fiber loop, the resonance frequency of the ESOP is shifted, but the resonance frequency difference does not take place between the CW and CCW light(Fig.2(b)). The Faraday effect with the twist induces the difference.

The variation of the bias according to the direction of the magnetic field ,for example by the change of the vehicle direction, can be regarded as the drift. It is expressed as

$$\delta = \frac{8\zeta_0}{\Delta\beta}\sqrt{2AW} \tag{7}$$

Equation (7) shows that the drift is inversely proportional to the fiber birefringence and proportional to the square root of the power spectrum component of the twist ,whose period is just equal to one turn of the fiber loop. This result is similar to that previously obtained for the interferometric fiber gyroscope[6,7].

Figure 3 shows the drift in the rotation rate calculated, by Eq.(7) with the relation between the sagnac phase shift and the rotation rate, as a function of the fiber birefringence. Figure 4 shows that as a function of the special twist component. The parameters used for the calculations are tabulated in Table 1. Typical drift required for the aircraft navigation is about 10^{-7}[rad/sec].

Consequently it has been found that the drift caused by the faraday effect can be supressed by use of the high birefringence fiber and decreasing the special twist component of the period just equal to one turn of the fiber loop.

4. Conclution

A theoretical investigation has been presented for the drift caused by the Faraday effect in OPRG. We have shown through it the ways of supressing the drift.

References

[1] K. Iwatsuki, K. Hotate and M. Higashiguchi; Appl.Opt.,23, 21,p.3916(1984)
[2] ---; J.Light wave Technol.LT-4,6,p.645(1986)
[3] ---; Appl.Opt.,25,15,p2606(1986)
[4] ---; Appl.Opt.,25,23,p.4448(1986)
[5] K. Bohm, K. Petermann and E. Weidel; Opt.Lett.,7,4,p.180
[6] K. Hotate and K. Tabe; Appl.Opt.,25,7,p1086(1986)
[7] ---; OFS'86, TOKYO, p.189

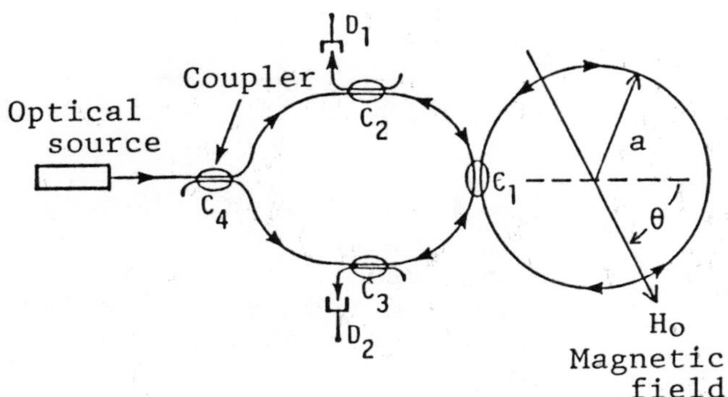

Fig.1 Basic setup of the OPRG using fiber ring resonator

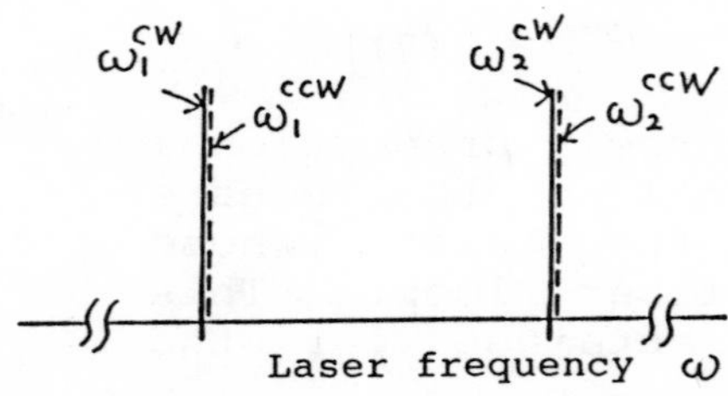

(a) Without Faraday effect nor twist

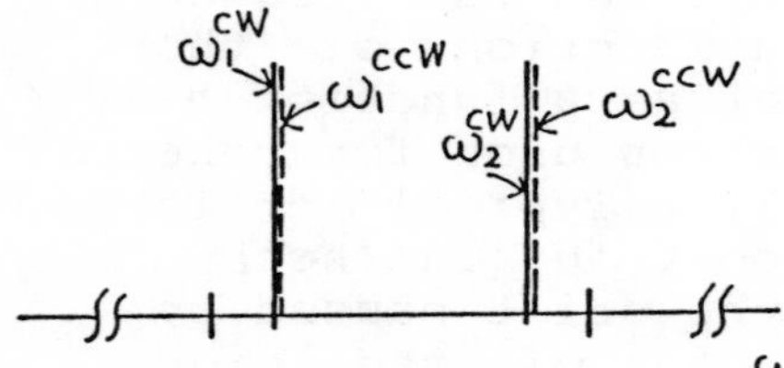

(b) With only twist

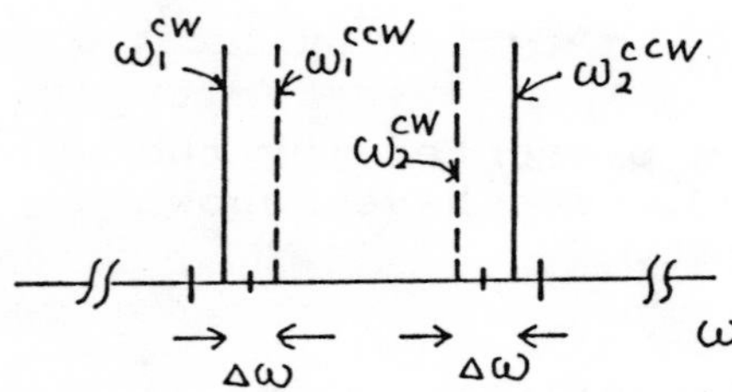

(c) With Faraday effect and twist

Fig,2 Resonance frequency difference between the CW and CCW light with the Faraday effect and the fiber twist. $\Delta\omega$ corresponds to $\Delta\psi$ in Eq.(5)

Faraday rotation	ζ_0	0.0001(rad/m)
(corresponding to the earth's magnetism)		
Radius	a	0.05(m)
Wave length	λ	0.8(μm)

Table,1 Parameters used for the calculations.

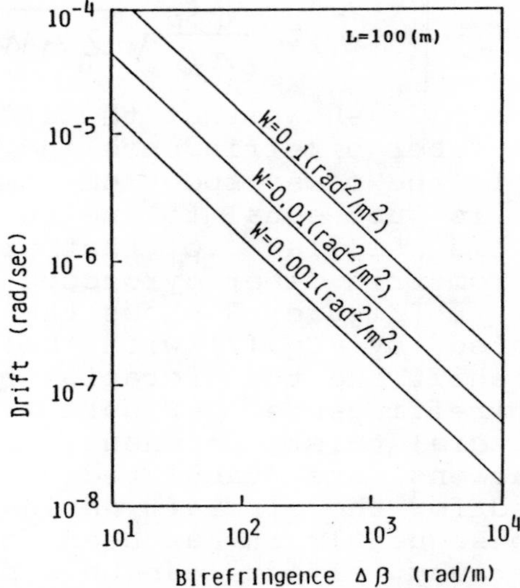

Fig.3 Relationship between the drift and the fiber birefringence

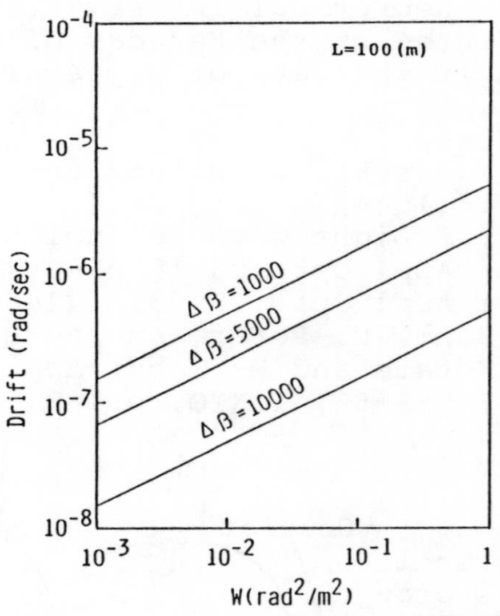

Fig.4 Relationship between the drift and the twist power-spectrum component whose period is just equal to one turn of the sensing loop

Evaluation of Polarization Maintaining Fiber Resonator for Rotation Sensing Applications

G.A. Sanders, N. Demma, G.F. Rouse, and R.B. Smith
Honeywell Systems and Research Center, Minneapolis, MN 55418
(612) 782-7722

Introduction

Fiber-optic ring resonators have been investigated as one class of fiber-optic gyros, primarily based on the potential for reduced fiber length, increased lnearity and dynamic range, as well as reduced sensitivity to thermal transient gradients[1-4]. Single mode fiber resonators have been fabricated and tested for rotation sensing applications[1-3]. These devices require polarization-controlling (PC) loops[5] to properly match incident light into one of the two resonant polarization eigenstates of the ring resonator. In order to obtain low drift performance, it has been pointed[4,6] out that tight tolerances on the PC's must be maintained. Since the state of polarization in the fiber is sensitive to environmental factors, especially temperature, active control of the PC's has been suggested[1,4]. One attractive alternative for achieving the required polarization control is the use of high birefringence polarization maintaining fiber for the resonator loop[1,2,4], eliminating the need for polarization controllers.

In this paper we report our evaluation of the properties of a PM fiber resonator and discuss some of the key issues facing its use in fiber resonator gyroscopes. Since the passive resonator gyro relies on precise measurement (to better than 1 part in 10^6) of the resonance center frequencies for both clockwise (CW) and counterclockwise (CCW) directions, we have examined the resonant mode structure of the PM resonator. In addition the PM resonator has been incorporated into an experimental closed-loop bench-top-gyro system for preliminary evaluation as a rotation sensor.

Experimental Arrangement and Results

Figure 1 shows the experimental arrangement used to measure the resonance mode structure of the PM resonator. The resonator consists of a 3 meter loop of 0.63 micron wavelength York polarization maintaining fiber and an input/output directional coupler for injecting light into and tapping light out of the resonator ring. Most of the resonator fiber is coiled around a 4 inch diameter cylinder. A segment of the resonator loop is wrapped around a piezoelectric (PZT) cylinder for varying the resonator pathlength.

Light from a single frequency HeNe laser is polarized by a high extinction polarizer which is aligned with one of the principal axes (say the x-axis) of the fiber. Although the extinction ratio of the polarizer is >53dB our measurements indicate that the y-axis excitation is actually limited at 40-43dB due to input lens birefringence and the fact that the fiber modes are not strictly linearly polarized. The output light, and in particular the resonance dips, are sensed by a photodetector and recorded on a chart recorder. Figure 2 shows the output as the resonator free spectral range of 67 MHz is scanned by applying a sawtooth drive voltage to the PZT cylinder. The various curves depict the resonator output as the coil is heated slightly, thereby changing the total loop birefringence. The input polarization, the lead temperature and the coupler temperature were constant over the duration of the measurements.

As shown in Figure 2 <u>two</u> resonance dips are observed, the main intended resonance as well as a small dip marked by an arrow. The two dips are consequences of the two ring resonances, which correspond to nearly x and y polarized light in the ring. The main dip is maximized by adjusting the input polarization, with the intention of minimizing the small dip. The main dip measures a finesse of 50 giving a linewidth of 1.3 MHz. (The narrow linewidth of the HeNe laser does not contribute to this measured linewidth). The depth of the main dip is nearly 98% indicating a good match between the amount of coupling and the internal loss in the ring.

The small dip exists in part because of imperfect polarization extinction at the input lead, but also because of any polarization cross-coupling in the coupler or elsewhere in the ring. As shown in Figure 2 the two resonance dips drift with respect to each other as the loop temperature changes. This is a consequence of the temperature dependence of the birefringence in the PM fiber (an effect which does not present itself in round single-mode fiber). This effect is unavoidable for the thermal-stress-induced high birefringence of PM fiber. Over a total temperature range of less than 1°C the small dip moves completely across the free spectral range. A particular concern is that the small dip of the "other" polarization, when crossing the main dip, causes significant distortion and lineshape asymmetries of the observed main resonance. Although other experiments conducted in our lab suggest that the asymmetries are the same for light propagation in the other direction, this asymmetry poses significant problems for rotation sensing since the line center is difficult to define and measure accurately. Only with detailed analysis can this effect be quantified, although it is clearly desirable to avoid this effect during gyro operation.

To obtain a preliminary indication of gyro performance, the same resonator was employed in a bulk-optic configuration to simulate closed-loop gyro performance[7]. In this system, shown in Figure 3, the laser light at frequency f_o is split into two beams each of which are frequency-shifted by acousto-optic frequency shifters (A/O) prior to entering the resonator. The resonator pathlength is adjusted by means of a reference control loop such that the CCW resonance frequency is matched to the CCW input beam frequency f_o+f_1. The CW input frequency f_o+f_2 is then adjusted to matched the CW resonance frequency by means of a secondary servo driving a voltage controlled oscillator (VCO). In order to obtain a discriminant for sensing resonance a standard AC detection scheme is used whereby the input beams are phase modulated (PM) at frequencies[8] f_m and f_n and subsequently demodulated in phase sensitive demodulators, PSD1 and PSD2, respectively. In the presence of rotation, the resulting difference frequency between input beams $\Delta f=f_2-f_1$ is a linear measure of rotation rate.

To measure the gyro output versus rotation rate the experimental assembly was mounted on a rate table. The difference frequency Δf, obtained electronically by mixing the drive signals to the A/O modulators, was then recorded in 100 ms sample intervals by a computer data acquisition station. The gyro output was compared with that of a Honeywell GG1342 ring laser gyro (RLG) which was mounted on the same table. Data from both the fiber resonator gyro and the RLG were recorded simultaneously as the rate table was slowly accelerated through a ±100 deg/sec range of rotation. A total of 35 data points were recorded. Figure 4 shows the resulting frequency output from the fiber gyro (vertical axis), versus the calibrated output of the RLG (horizontal axis). The RLG output was known to be linear and stable to better than 10 ppm, which was beyond the accuracy of these experiments. As shown in the figure the gyro output exhibited no visible non-linearities. Preliminary analyses show better than 0.5% linearity. However, improved mechanical stability of our setup should improve this result. In addition, preliminary noise and short bias stability investigations were conducted under laboratory temperature conditions. A random noise coefficient of 0.1 deg/root-hour was obtained, which was within a factor of six of the photon-noise limit for this system. Short-term bias stabilities of better than 50°/hr over 1/2 hr were observed.

Discussion & Conclusions

We have presented results from experiments with a passive all-fiber ring resonator fabricated of polarization maintaining fiber. A HeNe laser and bulk optical components outside the ring were used. We believe that these results are the most extensive reported to date for a PM fiber resonator. Characteristic of such a PM resonator is that the two resonance frequencies, for each the two polarizations within the ring, drift with respect to each other with temperature. The unwanted resonance, when excited by small amounts of cross-coupled light, leads to significant line-shape distortion and rate errors whenever the two resonances coincide. These must be avoided, possibly by the use of single-polarization fiber[1,3] or other polarizers within the ring path to suppress one resonance. Alternative techniques are under investigation. When operated as a closed loop gyro a linearity of better than 0.5% up to ± 100 deg/sec was obtained, an encouraging preliminary result toward obtaining useful scale factor performance for a range of applications. A random noise coefficient of 0.1 deg/root-hr was obtained, and a bias stability of 50°/hr over 30 min. was observed. Based on these results as well as others[1,2], we believe that the primary issue for fiber resonator gryos is bias stability rather than short-term noise. The above errors are attributable primarily to imperfect polarization control and mechanical instabilities in the bulk optic system used. These should be greatly improved by an all guided wave system.

References

1.) R.E. Meyer et al., Opt. Lett. 8, 644 (1983).
2.) W. Schroder, et al., in SPIE proc. Vol. 719, Fib. Opt. Gyros:Ten Anniv. 162-168 (1986).
3.) R. Carroll et al., in SPIE Proc. Vol. 719, Fib. Opt. Gyros:Ten Anniv. 169-177 (1986).
4.) K. Iwatsuki et al., Appl. Opt. 25, 25, 2606 (1986).
5.) H.C. Lefevre, Electron Lett. 16, 778 (1980).
6.) B. Lamouroux et al., Opt. Lett. 7, 8, 391-393 (1982).
7.) G. Sanders et al, Opt. Lett. 6, 11 (1981).
8.) K. Iwatsuki et al., Appl. Opt. 23, 3916 (1984).

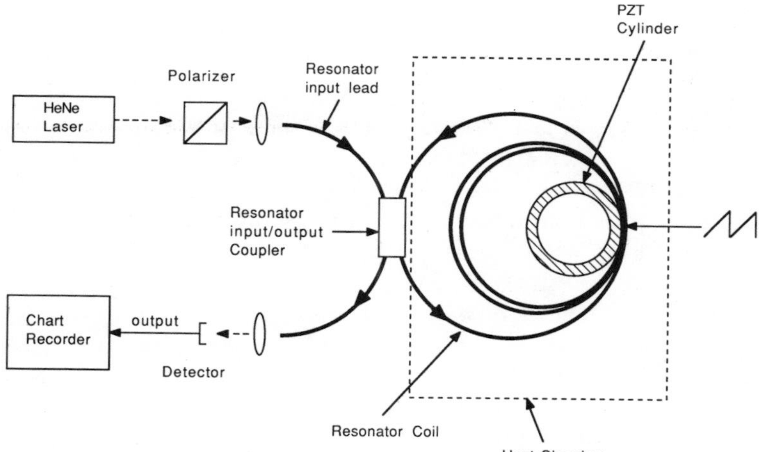

Figure 1 - Schematic of Experiment used to measure mode structure of PM Fiber Resonator

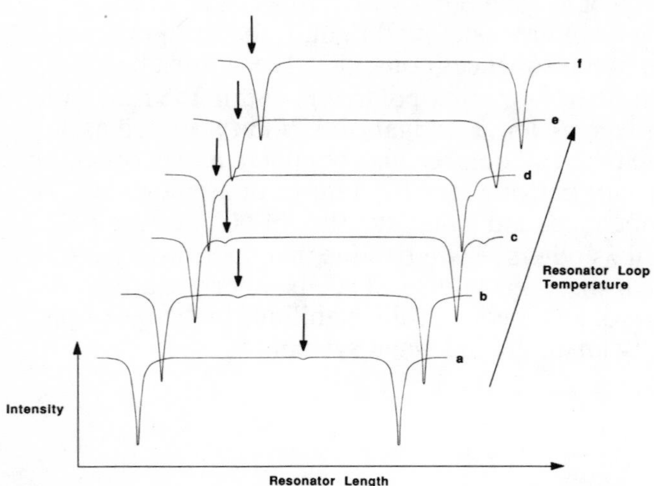

Figure 2 - Output Intensity versus Resonator Length as Temperature increases

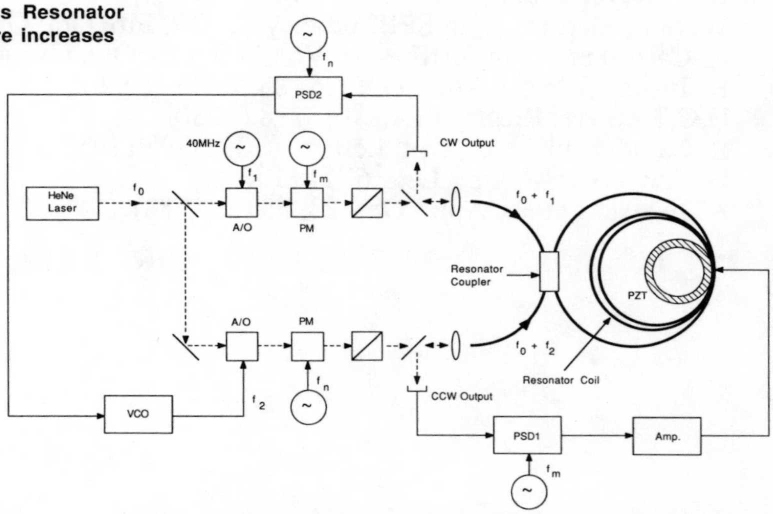

Figure 3 - Setup for Evaluating PM Fiber Resonator as a Rotation Sensor

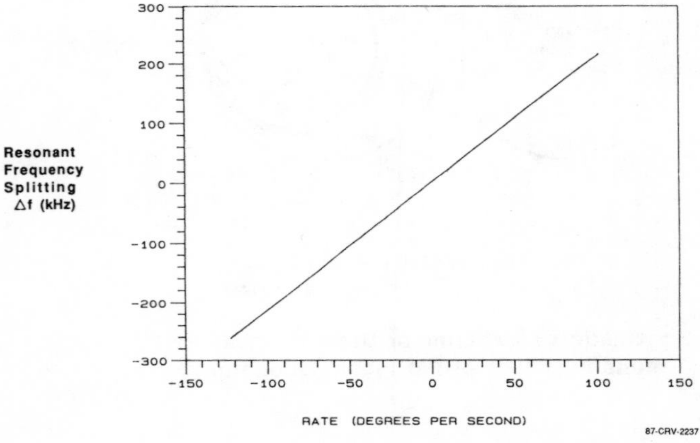

Figure 4 - Resonator Frequency Splitting versus Measured Rotation Rate

FRIDAY, JANUARY 29, 1988

MEETING ROOM 2-4-6

10:30 AM–12:15 PM

FCC1–6

STRAIN, PRESSURE, AND FLOW SENSORS

Kazuo Hotate, University of Tokyo, Japan, *Presider*

OPTICAL FIBER SENSORS FOR COMPOSITE STRUCTURES

James R. Dunphy and Gerald Meltz
United Technologies Research Center
Mail Stop 92
Silver Lane
East Hartford, CT 06108
(203)727-7006

S U M M A R Y

The compatibility of optical fiber sensors with composite materials enables one to install internal monitoring devices prior to curing. Proper installation of the minute, but strong, glass optical fibers within the laminate can provide a void-free component with minimal disturbance in the layup of the reinforcement filaments. We believe that the utility of such an installation will greatly improve the quality, as well as help extend the lifetime, of the composite structure by providing quantitative, realtime measures of its stress state. Furthermore, such protected, subsurface sensors can be used to facilitate measurements for active structural control. Two specific devices will be discussed for potential application to curing process control, residual stress assessment and in-service health monitoring. Both sensors use dual mode fibers. That is, measurements are conducted on the differential phase shifts between two modes. In essence, the sensors perform as linearly distributed interferometers. The relative phase change in response to a measurand is manifest as a crosstalk change in the twin-core fiber case and a polarization change in the birefringent fiber case. Furthermore, we introduce multiple wavelength operation of these dual mode fibers in order to resolve two parameters. For instance, the twin-core fiber can measure temperature and strain simultaneously, while the birefringent fiber can provide a principal 2-D stress analysis.

The principles involved in application of the multiwavelength birefringent sensor are closely analogous to classical photoelastic methods of stress analysis, only on a microscopic scale. During our experiments, a polarimeter is used to measure the change in polarization of light waves in an optical fiber due to a strain perturbation produced by intense stress waves propagating normal to the fiber axis. The fiber is buried in the composite specimen with its principal axes aligned with the major component of the compressive stress wave. As the stress wave passes, additional birefringence is induced. This birefringence change is proportional to the difference between σ_x and σ_y, the principal transverse stresses. Sensor operation in the stress difference mode is initiated by illuminating it at 45 degrees to the transverse, inherent fiber axes. On the other hand, if a second wavelength illuminates the sensor along one of its principal axes and a Mach-Zehnder or homodyne detection system is used to measure the relative phase delay along the sensor, then a weighted sum of σ_x and σ_y is measured. The two independent measurements provide a detailed principal stress analysis on a fine scale within the composite sample. Results from recent feasibility studies in the stress difference mode will be discussed to identify generic signals and to show stress wave dispersion within composite materials. The stress disturbances were created with pulsed CO_2 laser irradiation of the sample surface and were recorded with a few nanoseconds response time.

The twin-core fiber sensor is quite different from the birefringent device. For instance, the fiber has two, matched single mode cores which are very closely spaced. When one core is illuminated, both symmetric and antisymmetric modes of the fiber are equally excited and light couples back and forth betweeen the cores as it propagates along the fiber. Complete power transfer takes place over a distance defined as the beat length λ_b of the dual mode interference pattern which is linearly distributed down the sensor. The variation in light intensity in each core in a length L is a periodic function of the beat phase $\phi_b = \pi\lambda_b/L$. In general both L and λ_b vary with temperature and strain. The net effect of a measurand is to change the beat phase and the core contrast $Q = (P_1 - P_2)/(P_1 + P_2)$. Q is a cosinusoidal function of ϕ_b. Temperature and strain can be measured simultaneously by detecting the response in Q at two wavelengths. The change in beat phase due to an axial strain (ϵ) and a temperature variation (T) is

$$\delta\phi_b = A\ (T,\lambda)\ T + B(T,\lambda)\epsilon$$

where the coefficients A and B contain the device sensitivities to temperature, strain and wavelength (λ). For instance, A includes the effects of thermo-optic refractive index variations and stress-optic changes arising from the thermal strain of the core. Photoelastic effects are contained in B, accounting for both index and dimensional changes due to the applied strain. To a good approximation, the strain sensitivity is independent of temperature. Therefore, measurements of Q at two wavelengths yields sufficient information to determine both T and ϵ. The calibration constants A and B can be calculated or measured in the laboratory. The twin-core device has been used to measure several thousand microstrains at roughly 1200 degrees Fahrenheit and is clearly compatible with the curing requirements of reinforced polymer composites. This paper will review demonstrations in which the twin-core sensor monitored composite panel stresses as well as fatigue and fracture which developed in a specimen with a cyclically loaded bolt hole.

We conclude that the compatibility and versatility of optical fiber sensors in application to composite material diagnostic problems will lead to many practical nondestructive evaluation methods. These subsurface devices will support the advancement of fatigue and fracture models and curing process control. Furthermore, as the optical fiber technology is pushed into temperatures approaching 2000 degrees F., glass and ceramic matrix materials can be diagnosed. Finally, development of smart structures with integral sensors and actuators for active deflection, vibration and thermal control is imminent.

Elliptical Core Two-Mode Fiber Strain Gauge With Heterodyne Detection

J. N. Blake, Q. Li, and B. Y. Kim

Edward L. Ginzton Laboratory, W. W. Hansen Laboratories of Physics
Stanford University, Stanford, California 94305

Abstract

Detection of strains using modal interferometry in elliptical core two-mode optical fibers is becoming practical. We report two methods of heterodyning this type of strain sensor in order to obtain an rf output signal whose phase is linearly proportional to the axial strain applied to the fiber. Heterodyning the strain gauge greatly improves its sensitivity and overcomes the signal fading problem encountered when directly analyzing the interference pattern.

Strain Detection In Two-Mode Elliptical Core Fibers

Modal interferometry has been employed with limited success for detection of hydrostatic pressure using circular core two-mode fiber[1] and for detection of strain using two-core fiber.[2] We have recently demonstrated the use of elliptical core two-mode fiber for strain detection.[3] Elliptical core fibers have the property that they propagate only the fundamental LP_{01} mode and one lobe orientation of the second order LP_{11} mode over a large region of the optical spectrum.[4,5] Since the LP_{11} mode pattern is uniquely defined and stable, modal interferometry becomes practical.[5] Also, the elliptical geometry of the core introduces birefringence and the fiber can be polarization maintaining.

The principle of strain detection is as follows. Light is launched with approximately equal intensities into the LP_{01} and the LP_{11} modes. The two eigenpolarizations, along the major and minor axes of the core ellipse, are decoupled so we can consider them separately. At the fiber output, light in the symmetric LP_{01} mode interferes with light in the antisymmetric LP_{11} mode yielding a far-field radiation pattern whose intensity distribution is a function of the phase difference between the two modes. Longitudinal strains applied to the fiber change the relative phase between the two modes.[3,6] Thus light in the far-field shifts from side to side sinusoidally with applied strain. Consequently the light intensity falling on a detector offset from the fiber axis varies sinusoidally with applied strain. Alternative methods for detecting the relative phase between the two modes are described in reference 5.

Synthetic Heterodyning

The first method we demonstrate for heterodyning this strain sensor is akin to the synthetic heterodyning technique first applied to the fiber-optic gyroscope.[7] Figure 1 shows the system. Light is launched into the two-mode fiber (provided by the Andrew Corporation) with approximately equal intensities in each of the two spatial modes. The polarizers at the input and output of the fiber ensure that only one eigenpolarization is operated at a time. A section of the fiber is wound around a cylindrical PZT phase modulator. This provides a modulation of the relative phase between the LP_{01} and LP_{11} modes at the fiber output. The detector current, I_d, is proportional to the light intensity falling on the detector and is given by

$$I_d \propto I_0[1 + K cos(\phi_0 + m cos 2\pi f_m t)] \tag{1}$$

where I_0 represents the total light in the fiber, K is the electrical fringe visibility dependent on the mode content in the fiber and the area and location of the detector, f_m is the signal (26 KHz) applied to the PZT, m is the depth of modulation of the relative phase between the two modes, and ϕ_0 is the dc value of the relative phase between the two modes (dependent on the strain present) to be sensed. Setting the PZT drive voltage such that $m = 2.8$ radians and passing this detector current through the circuit shown[7] in fig. 1. yields the output current

$$I_{out} \propto cos(4\pi f_m t + \phi_0). \tag{2}$$

Comparing this signal with a frequency doubled PZT drive signal yields ϕ_0 and hence the applied strain. Figure 2 shows a plot of values of ϕ_0 measured in this way versus fiber elongation for both eigenpolarizations. Different amplitudes were used for the PZT drive signal to obtain $m = 2.8$ radians for each eigenpolarization.

Frequency Shifting

The second method for providing a heterodyned output, depicted in figure 3, makes use of the two-mode fiber acousto-optic frequency shifter.[8,9] Light from an argon laser ($\lambda = 488$nm) is launched into a two-mode fiber (provided by the Polaroid Corporation) preferably in the fundamental LP_{01} mode. A short section of the input end of the fiber is tightly wound around a small radius to strip off any second order LP_{11} mode inadvertently excited at the input. The frequency shifteris set to operate at 50% coupling efficiency and so converts one-half of the light to the second order LP_{11} mode. Thus after the frequency shifter, one-half of the light is in the fundamental mode unshifted in frequency, and one-half of the light is in the second order mode down shifted in frequency, in this case by 9.2 MHz. The light then enters the strain sensing region where externally applied axial strains induce a differential phase shift between the first and second order modes. The differential phase shift induced between the modes is different for the two eigenpolarizations of the elliptical core two-mode fiber. A polarization analyzer at the output of the two-mode fiber allows detection of either eigenpolarization. Both eigenpolarizations will generally be present since the LP_{11}mode stripper scrambles the polarization state of the light.

Since the two spatial modes have differing optical frequencies, the phase shift between the modes induced by axial strains is detected as a phase shift in the rf signal in the detector. The detector current, I_d, is given by

$$I_d \propto I_0 [1 + K cos(2\pi f_m t + \phi_0)] \tag{3}$$

where I_0 and K are as defined before and f_m is the frequency of the 9.2 MHz signal applied to the frequency shifter. Now ϕ_0 can be directly measured by comparing the phase of the detector current signal with that of the signal applied to the frequency shifter.

Figure 4 shows plots of the rf phase shift versus fiber elongation for the two eigenpolarizations. As expected, the responses of the two heterodyned interferometers are linear and the slopes match the values expected from previous dc strain sensitivity measurements.[6]

Conclusion

In conclusion we have demonstrated two methods for heterodyning the two-mode fiber strain gauge to obtain output rf signals whose phase is linearly proportional to the fiber elongation in the strain sensing region. Each of these methods provides a practical way of obtaining high sensitivity and eliminating the signal fading problem for those applications that allow for an active element to be present near the sensing region.

This research was supported by Litton Systems Inc.

References

1. M. R. Layton and J. A. Bucaro, Appl. Opt. Vol. 18, 464 (1979)

2. G. Meltz, J. R. Dunphy, W. W. Morey, and E. Snitzer, Appl. Opt. 22, 3 (1983)

3. S. Y. Huang, J. N. Blake, B. Y. Kim, and H. J. Shaw, "Highly Elliptical Core Two-Mode Fiber Strain Gauge" submitted to Opt. Lett.

4. A. W. Snyder and X. H. Zheng, J. Opt. Soc. Am. A /vol 3, 600 (1986)

5. B. Y. Kim, J. N. Blake, S. Y. Huang, and H. J. Shaw, "Use Of Highly Elliptical Core Fibers For Two-Mode Fiber Devices" to be published in Opt. Lett. Sept. (1987)

6. J. N. Blake, S. Y. Huang, B. Y. Kim, and H. J. Shaw, "Strain Effects On Highly Elliptical Core Two-Mode Fibers" to be published in Opt. Lett. Sept. (1987)

7. B. Y. Kim and H. J. Shaw, Opt. Lett. 9, 378 (1984)

8. B. Y. Kim, J. N. Blake, H. E. Engan and H. J. Shaw, Opt. Lett. 11, 389, (1986)

9. J. N. Blake, B. Y. Kim, H. E. Engan and H. J. Shaw, Proceedings of the SPIE, 719, (1986)

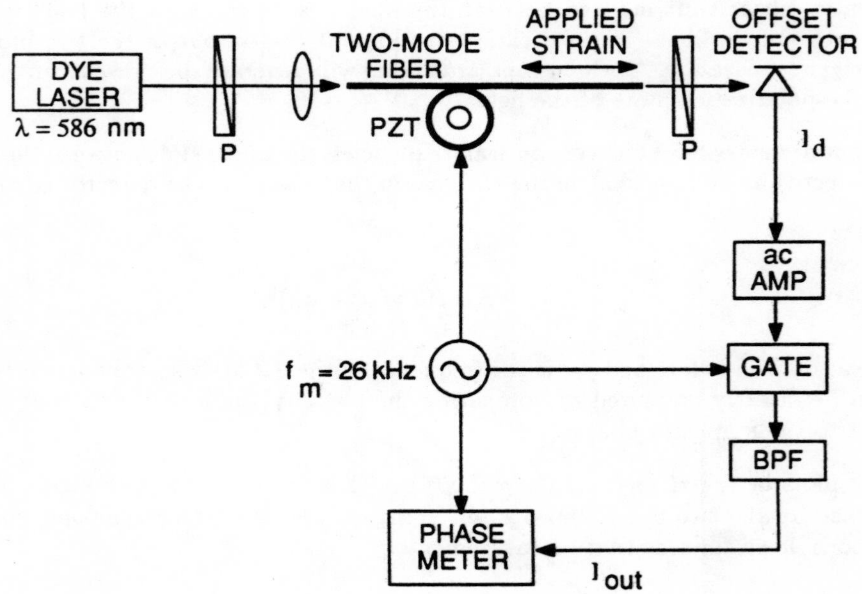

Fig. 1. Synthetic heterodyned strain gauge.

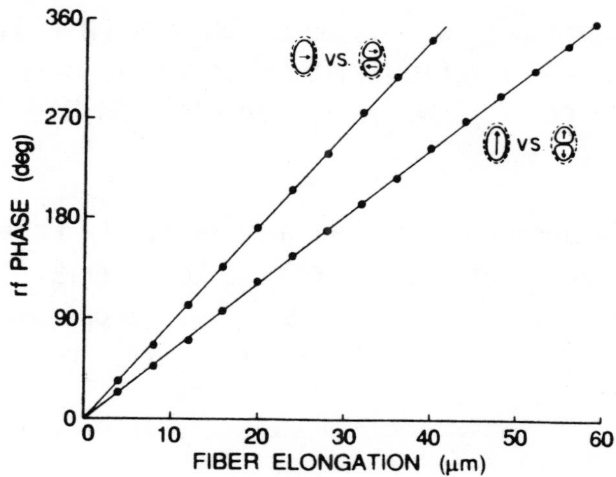

Fig. 2. Response of synthetic heterodyned strain gauge to Andrew fiber elongations

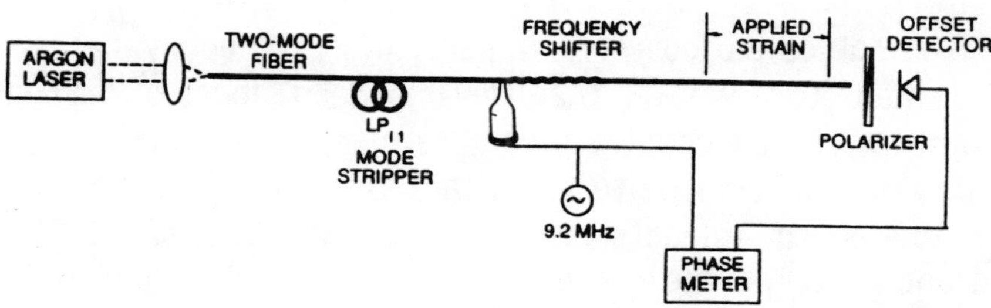

Fig. 3. Heterodyned strain gauge using two-mode acousto-optic frequency shifter.

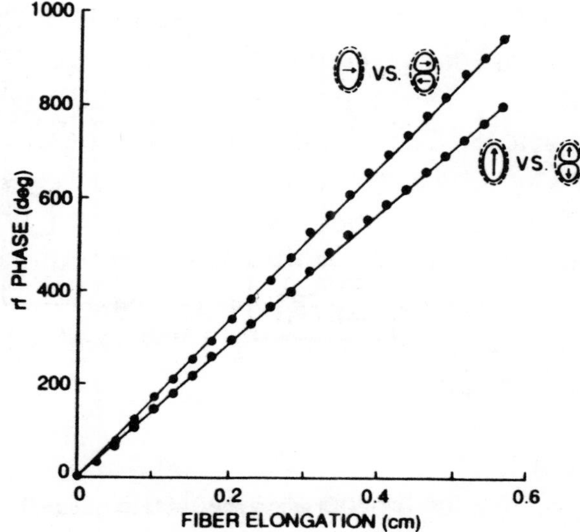

Fig. 4. Response of frequency shift heterodyned strain gauge to Polaroid fiber elongations

Development of a Structurally Imbedded Fiber Optic Impact Damage Detection System for Composite Materials

R. M. Measures, N. D. W. Glossop, J. Lymer and R. C. Tennyson
University of Toronto Institute for Aerospace Studies
4925 Dufferin St., Downsview, Ontario, M3H-5T6, Canada

Abstact

As part of a program to develop a structurally imbedded fiber optic impact damage detection system for composite material structures we have undertaken a careful study of the sensitivity of the optical fibers to to their orientation with respect to the adjacent plies and their depth in regard to the impact surface location.

Introduction

Composite materials, such as carbon/epoxy or Kevlar/epoxy, have values of strength to weight ratio, stiffness and resistance to fatigue and corrosion that make them very desirable for the aerospace industry. Unfortunately, the very nature of these fiber reinforced polymers endows them with anisotropic properties. In particular, the epoxy matrix is at least an order of magnitude weaker than the imbedded fibers of carbon or Kevlar. Disbonding between successive plys - termed " Delamination ", represents the Achilles heel of composite materials. Delaminations can be induced by relatively low energy impacts, such as dropping a tool or a collision with a bird, and once started the region of disbonding will grow with continued stress cycling.

Current non-destructive techniques including: (i) visual inspection, (ii) X-ray examination, (iii) holographic interferometry and (iv) ultrasonic probing (C-scan), cannot always be used for detecting damage to aircraft structural components made of composite materials. This is certainly the case for aircraft leading edges fitted with deicing rubber boots.

The development of a composite material damage detection technique based on a grid of optical fibers imbedded within the structural component at the time of manufacture could be of considerable benefit in such instances. This grid of optical fibers would be interrogated by a source of light and the disruption of its integrity used to determine the location and extent of the damage sustained as a result of an impact. We have undertaken the development of such a **Structurally Imbedded Fiber Optic Impact Damage Detection, SIFOIDD** system.

Experimental Program

The essential elements of a program devoted to the development of such a **SIFOIDD** system includes:

(i)Devising a treatment that would permit the optical fibers to have a strength commensurate with their role as damage detectors.

(ii)Perfecting the technology of imbedding optical fiber grids into the composite material structures at their time of manufacture.

(iii)Ascertaining the optimum orientation of the optical fibers with respect to the material fiber orientation in the adjacent plies, the number of optical fiber grids and their depth in regard to the impact surface.

(iv) Ensuring that the imbedded optical fiber grids do not have a detrimental influence on the resistance of the structure to delamination caused by impact or stress fatigue.

(v) Developing the technology of optically coupling large grids of optical fibers to a diagnostic port.

(vi) Constructing the appropriate interrogation unit, which would comprise a photodetection, data acquisition and computer processing system.

(vii) Creating the software for data reduction and interpretation.

The basic experimental setup for undertaking the preliminary phase of this research program is shown schematically in figure 1. A composite panel is manufactured containing a grid of specially treated imbedded optical fibers. The 632.8 nm light from a Helium-Neon laser is focussed onto the end of a fiber optic bundle that is connected to the imbedded optical array. The input optics is adjusted until each of the optical fibers is fairly uniformly illuminated. The composite panel is freely supported and the reference distribution of light transmitted through the optical grid recorded. The panel is then struck with a weight of known mass and measured velocity. Damage to the panel as indicated by fracture of the optical fibers is then compared with other methods of observation.

Experimental Results

Our initial work was undertaken on square panels of Kevlar/epoxy with a 0/90/90/0, ply configuration. Specially treated optical fibers were imbedded between various plies and with different orientations as indicated in figure 2. The translucent nature of these 4-ply panels made it possible pinpoint the position of fracture along each optical fiber from the bleeding of laser radiation that occurs at such locations. Furthermore, backlighting of the panel revealed the extent of any delamination that had arisen. This simplified approach has permitted us to perform an extensive analysis of the sensitivity to fracture of the optical fiber with respect to its orientation compared to the orientations of the two adjacent plies.

Our observations indicate that configuration-5, see figure 2, appears to be the optimum in regard to providing a reliable indicator of delamination. In this configuration the optical fiber is sandwiched orthogonaly between a pair of colinear plies. Fracture of the optical fibers seems to be associated with the cracking observed through colinear plies in the region of delamination. It should be noted that in most cases although the panels exhibit barely visible surface damage (even at the point of impact) their compressive strength has been found to be severely compromised by the internal damage. It is also found that optical fibers sandwiched diagonally between the two plies closest to the impact surface (configuration-3) are the least sensitive, while those sandwiched diagonally between the two plies that are furthest from the impact surface (configuration-9) are the most sensitive.

A postmortem of many panels has indicated that delamination usually occurs between the two plies furthest from the impact surface and that delamination almost never arises between colinear plies. We have also checked that the imbedded optical fiber grids do not have a detrimental influence on the compressive strength of the composite material and are currently studying its effect on the resistance of the structure to delamination caused by impact or stress fatigue.

Conclusions

In the course of developing a structurally imbedded fiber optic impact damage detection system for composite material panels we have determined that the optimum orientation and location for the optical fibers, in regard to providing a reliable indicator of delamination, is sandwiched orthogonaly between a pair of colinear plies. We have also shown that these optical fibers can detect damage that is barely visible on the impact surface but is sufficient to severely compromise the compressive strength of the panel.

Acknowledgements

This work is partly supported by the Natural Science and Engineering Research Council of Canada.

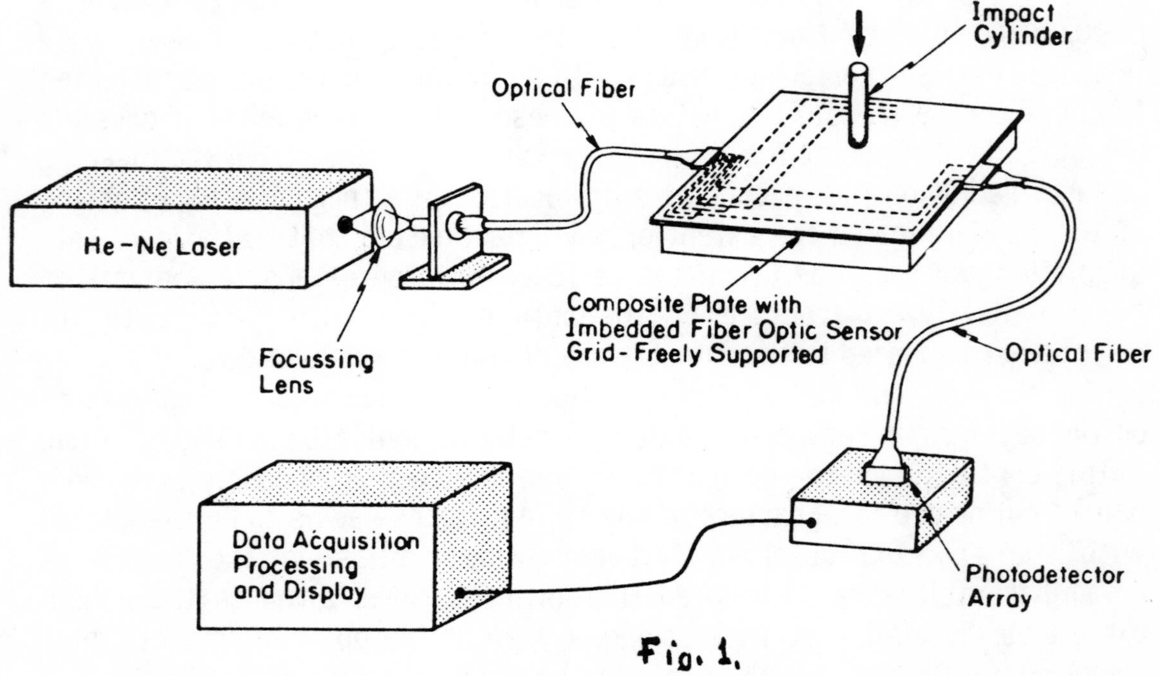

Fig. 1.

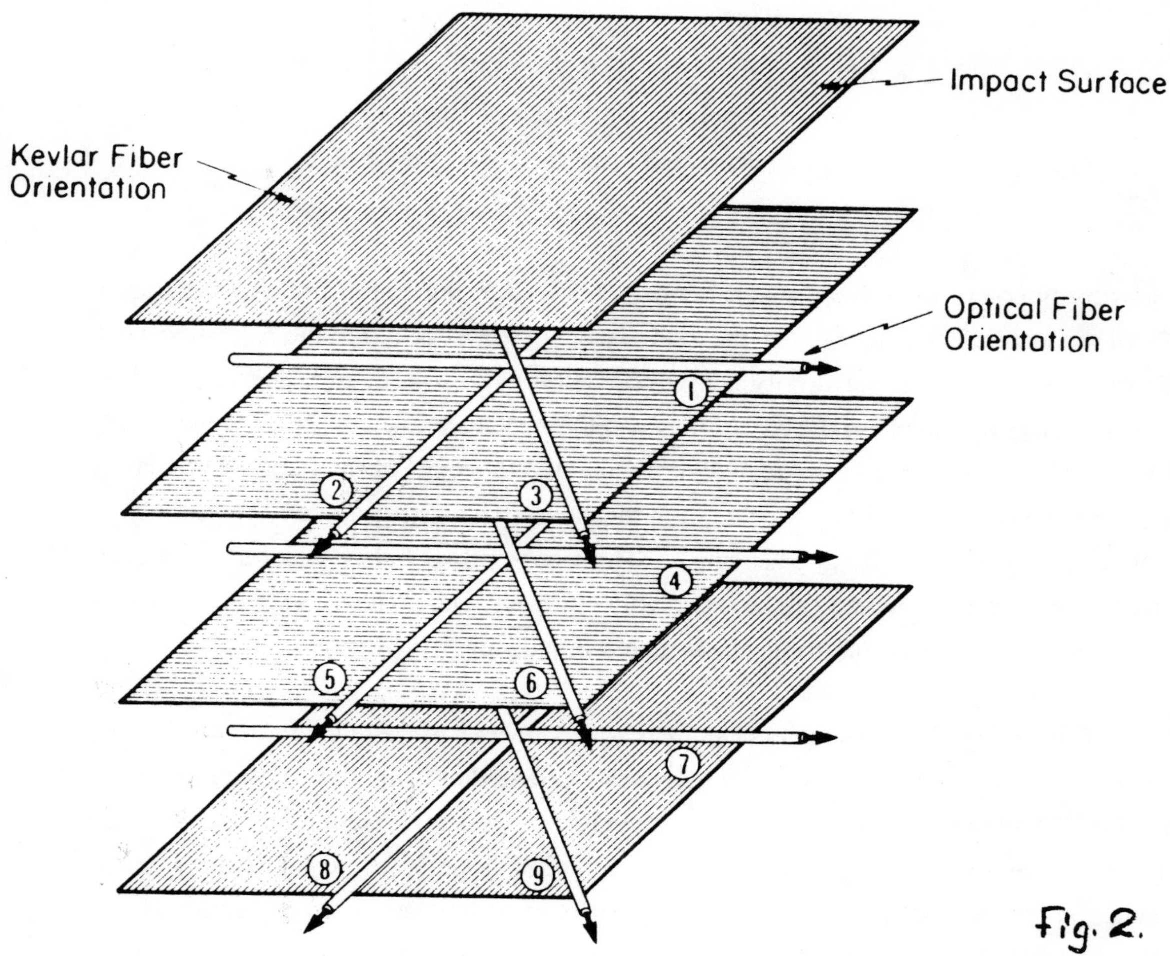
Impact Surface
Kevlar Fiber Orientation
Optical Fiber Orientation
1
2
3
4
5
6
7
8
9

Fig. 2.

HIGH SENSITIVITY FABRY-PEROT OPTICAL FIBER SENSOR FOR THE MEASUREMENT OF MECHANICAL FORCE

F. Maystre, P. Gannagé, R. Dändliker
Institute of Microtechnology, University of Neuchatel
Breguet 2, CH-2000 Neuchatel, Switzerland, +41-38-24 60 00

A high sensitivity polarimetric fiber sensor has been realized, using a high finesse fiber optical Fabry-Perot resonator. The frequency of the diode laser source is locked to the resonance peak of the Fabry-Perot, where the sensitivity is highest. Heterodyne polarization interferometry is then applied to measure the phase difference between the two eigenpolarizations, independently of their amplitudes. The sensor is insensitive to temperature changes of the Fabry-Perot cavity and to other perturbations that affect the length of the cavity. The sensitivity may be as much as 100 times larger than for a single-pass sensor. The fiber optical Fabry-Perot as a passive, lightweight sensor element, which can be made very compact. The application to the measurement of static force by induced birefringence is reported in detail.

High finesse fiber Fabry-Perot resonators can be produced from a piece of single-mode fiber by carefully polishing the fiber ends and gluing properly oriented chip size dielectric mirrors on both ends [1]. When the fiber, which constitutes the resonator cavity, is pressed between two parallel plates, the stress distribution in the core region is anisotropic and brings linear birefringence in the cavity [2]. The light that propagates in such a cavity can be separated into two linear eigenpolarizations, oriented along the two orthogonal axes of birefringence, which see sightly different indices of refraction, or optical length. One gets a dual Fabry-Perot cavity, implemented in parallel on a common length of fiber. In this paper, it will be shown how a high sensitivity static force sensor can be made from a fiber Fabry-Perot resonator by measuring the phase difference between the two eigenpolarizations at the output of the resonator. External perturbations affecting the length of the fiber or the wavelength of the source are common-mode perturbations and do not change the observed value of induced birefringence. The gain in sensitivity compared to a single-pass sensor is proportional to the finesse and can become more than 100.

Figure 1 shows the intensity and the phase transfer functions for the transmission of a birefringent Fabry-Perot resonator of finesse $F = 10$ as a function of the mean accumulated single pass phase $\phi_0 = (2\pi/c)\nu d n_0$, where c is the speed of light in vacuum, ν is the light frequency, d is the length of the resonator and n_0 is the mean refractive index of the two eigenmodes. The transmitted intensity consists of two identical Airy combs, separated by the phase difference $\Delta\phi = (2\pi/c)\nu d\Delta n$, where $\Delta n = n_1 - n_2$ is the difference of the refractive indices of the two eigenpolarizations. For a transverse force K applied to the uncoated fiber, the induced birefringence yields $d\Delta n = 4CK/\pi R$, where C is the stress-optic coefficient ($C = -3.7\times10^{-12}$ m^2/N for silica) and R is the fiber diameter [2]. The phase difference between the two transmitted eigenpolarizations is given by $\Delta\psi = \psi_1 - \psi_2 = 2\tan^{-1}[\rho\tan(\Delta\phi/2)]$, with $\rho \cong (1 + 2F/\pi)$ for a resonator finesse of $F > 1$ [3].

The fiber Fabry-Perot force sensor is based on the measurement of this phase difference $\Delta\psi$. Around the resonance, $\Delta\phi = m\pi$, $\Delta\psi$ depends linearly on $\Delta\phi$ and the slope, or sensitivity, is readily seen to be $\Delta\psi/\Delta\phi = \rho$ (for example $\rho = 7$ for $F = 10$), instead of $\Delta\psi/\Delta\phi = 1$ in the case of a single-pass sensor. However, it is important to note that the dynamic range of the described Fabry-Perot sensor is limited to the width of the resonance and that, because of the large amplitude variations near the resonance, it is imperative to measure the phase independently of the amplitude. The detection scheme which has been chosen can be divided into two parts. The first part consists of a feedback loop, which locks the laser frequency to the Fabry-Perot resonance peaks, where the sensitivity is maximum, and the second part measures the phase difference $\Delta\psi$ accurately using heterodyne detection.

The electronics for the stabilization of the laser frequency is shown in the upper part of Fig. 2. The temperature stabilized single-mode laser diode is frequency modulated at 1 kHz via the injection current over a range corresponding to a fraction of the resonance width. The error signal is proportional to the corresponding frequency component in the output intensity of the Fabry-Perot [4]. This error signal is extracted from the mean intensity of the two eigenmodes, detected by the photodiode D2 behind a polarizer P2 oriented at 45°. The frequency locking works properly as long as the two resonance peaks (Fig. 1) are separated by less than their width, which is about $\Delta\phi = \pm1.2/F$. This limits the dynamic range. At this point, the sensitivity $\Delta\psi/\Delta\phi$ is still about 85% of its maximum value for $\Delta\phi = m\pi$.

The heterodyne polarization interferometer, which allows to measure the phase difference $\Delta\psi$ of the two eigenpolarizations independently of their amplitudes, is shown in the lower part of Fig. 2. The light from the laser is split by the polarizing beam splitter PBS into two beams with orthogonal polarizations. Two bragg cells, M1 and M2, shift the optical frequency in the two arms by 40.0 MHz and 40.1 MHz, respectively, so that the beat frequency is 100 kHz, which is compatible with the electronic phasemeter used in the experiment. The two polarizations are recombined through the non-polarizing beam splitter BS. The photodiode D1, placed behind a polarizer P1 oriented at 45°, supplies the heterodyne reference signal. The complementary beam is launched into the fiber Fabry-Perot resonator FFP. Care is taken to ensure that the two orthogonal polarizations are aligned parallel and perpendicular to the principal axis of birefringence of the pressed fiber. At the output, the photodiode D2, placed behind a polarizer P2 oriented at 45°, detects the phase shifted heterodyne signal. The two signals are fed through a band pass filter to the electronic phasemeter, where the phase difference $\Delta\psi$ is determined by zero-crossing detection.

Measurements were made with a Fabry-Perot consisting of a 10 cm long single-mode fiber of low intrinsic birefringence with two gold coated mirrors of 76% reflectivity glued to both ends. The free spectral range of the resonator was 1 GHz and the finesse has been measured to be F = 10, which is in good agreement with the theoretical value for 76% mirrors. The fiber resonator is placed in a mechanical mounting that permits the application of a static transverse force on a short length of the fiber, where the coating has been removed. A bias load was applied in order to have the two resonances of the eigenpolarizations exactly superimposed (i.e. $\Delta\phi = \pi$). Around this working point, the measurements were made by adding and removing calibrated weights. The laser diode is optically isolated from the Fabry-Perot resonator and emits at 780 nm a single line of 20 MHz width (FWHM), which is about 5 times narrower than the Fabry-Perot resonance peaks. At the output of the optical isolator (OI), the light passes through a single-mode fiber (SMF), which acts as a spatial filter, before entering the heterodyne polarimeter.

The experimental results are shown in Fig. 3. The resolution for $\Delta\psi$ was limited to ±3° by instabilities of the single-mode fiber (SMF), acting as the spatial filter. Further improvements should bring the resolution closer to that of the phasemeter, which is 0.01°. The reproducibility was of the same order as the resolution. The experimental points follow closely the theoretical curve for a Fabry-Perot resonator of finesse F = 10 and a silica fiber of 125 μm outer diameter. The maximum slope is 2.1°/g, which is 7 times larger than the sensitivity of a single-pass sensor.

A novel high resolution polarimetric Fabry-Perot sensor concept, with a diode laser source and heterodyne phase detection of the transmitted light, has been successfully applied to measure static force by induced birefringence with increased sensitivity. In some cases, it might be more convenient to use the Fabry-Perot in reflection rather than in transmission, since this requires access to only one fiber end. However, in order to have enough reflected intensity at the resonances, the Fabry-Perot has to be asymmetric with 100% reflectivity at the far end and a limited finesse F, namely F = 40 for $I_{res} = 0.2\ I_{max}$.

References

[1] J. Stone, "Optical-fibre Fabry-Perot interferometer with finesse 300" Electron. Lett. **21**, 504-505, (1985).

[2] A. Bertholds, R. Dändliker, "High-resolution photoelastic pressure sensor using low-birefringence fiber", Appl. Opt. **25**, 340-343 (1986)

[3] R. Kist, W. Sohler, "Fiber optic spectrum analyser", Journ. of Lightwave Technol. **LT-1**, 105-109 (1983).

[4] J-L. Picqué and S. Roizen, " Frequency-controlled CW tunable GaAs laser", Appl. Phys. Lett., **27**, 340-342 (1975).

Figure captions

Fig. 1 Intensity and phase of the transmitted eigenpolarizations in a birefringent Fabry-Perot resonator as a function of the mean accumulated single pass phase.

Fig. 2 Experimental setup for the measurement of the accumulated birefringence in a Fabry-Perot resonator using heterodyne polarimetry. LD: laser diode, L1, L2, L3, L4: objective lenses, OI: optical isolator, SMF: single-mode fiber (spatial filter), PC: polarization controller, FFP: fiber Fabry-Perot.

Fig. 3 Measured phase difference $\Delta\psi$ of the transmitted eigenmodes versus applied transverse load. Solid line: theoretical curve for a Fabry-Perot of finesse F = 10 and a silica fiber of 125 μm outer diameter. Dashed line: single-pass sensor for comparison.

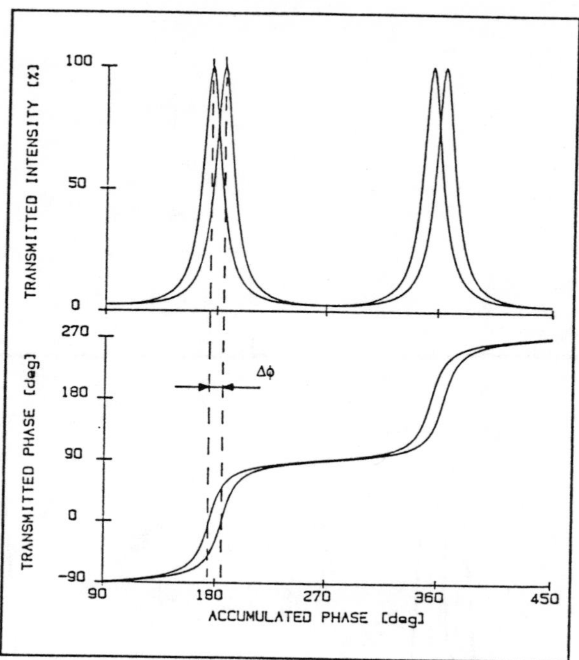

Fig. 1 Intensity and phase of the transmitted eigenpolarizations in a birefringent Fabry-Perot resonator as a function of the mean accumulated single pass phase.

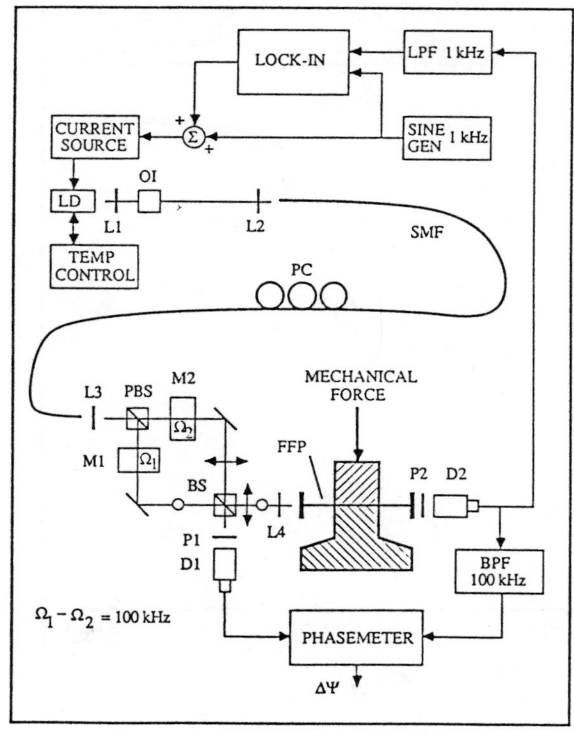

Fig. 2 Experimental setup for the measurement of the accumulated birefringence in a Fabry-Perot resonator using heterodyne polarimetry. LD: laser diode, L1, L2, L3, L4: objective lenses, OI: optical isolator, SMF: single-mode fiber (spatial filter), PC: polarization controller, FFP: fiber Fabry-Perot.

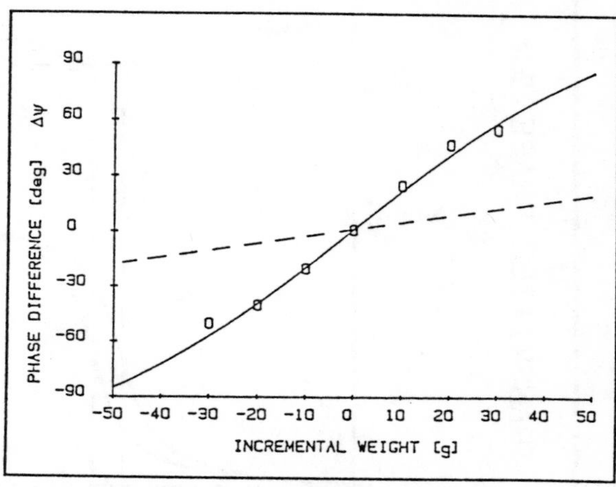

Fig. 3 Measured phase difference $\Delta\psi$ of the transmitted eigenmodes versus applied transverse load. Solid line: theoretical curve for a Fabry-Perot of finesse F = 10 and a silica fiber of 125 μm outer diameter. Dashed line: single-pass sensor for comparison.

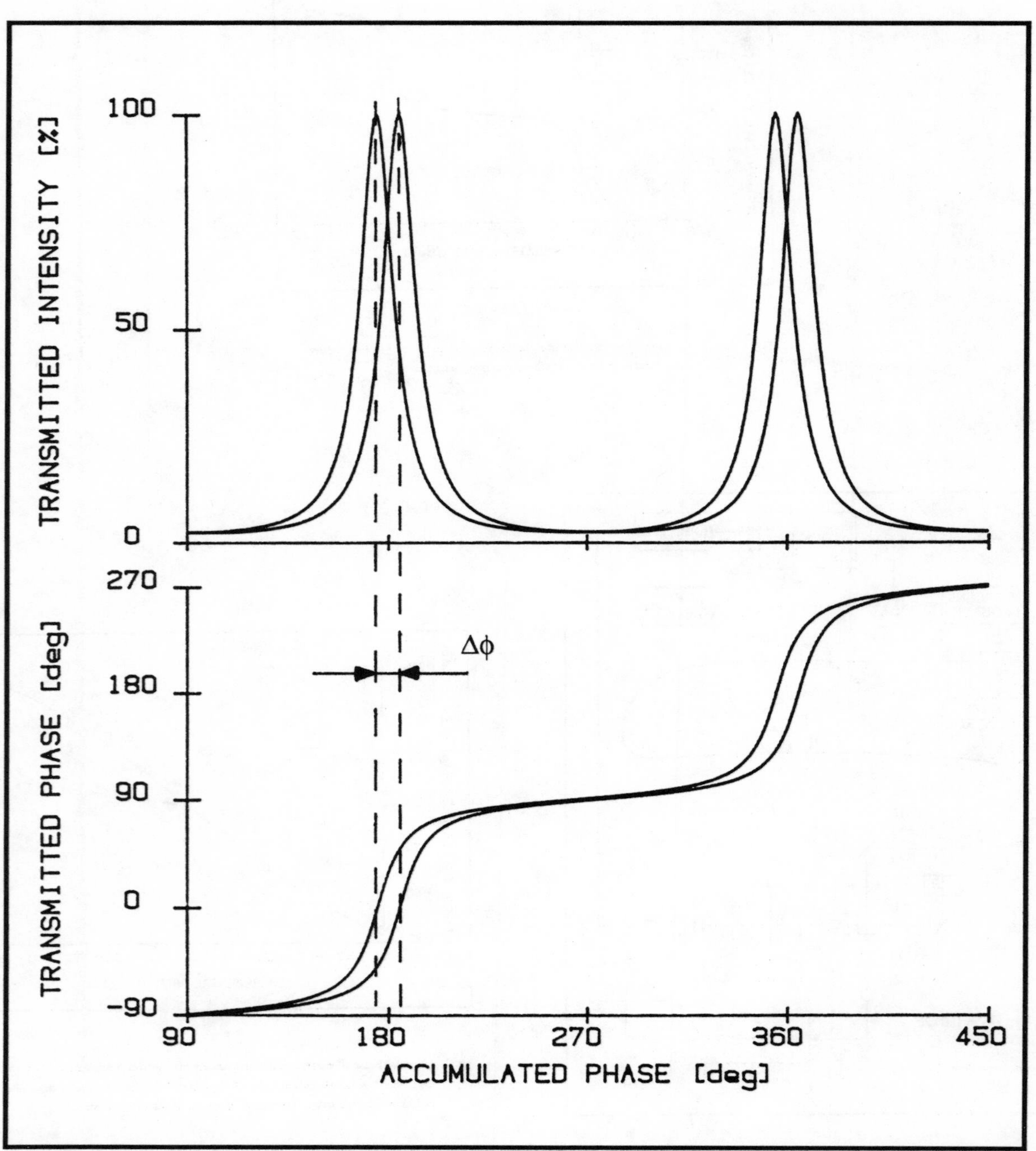

100
50
0
TRANSMITTED INTENSITY [%]
270
180
90
0
-90
TRANSMITTED PHASE [deg]
Δφ
90
180
270
360
450
ACCUMULATED PHASE [deg]

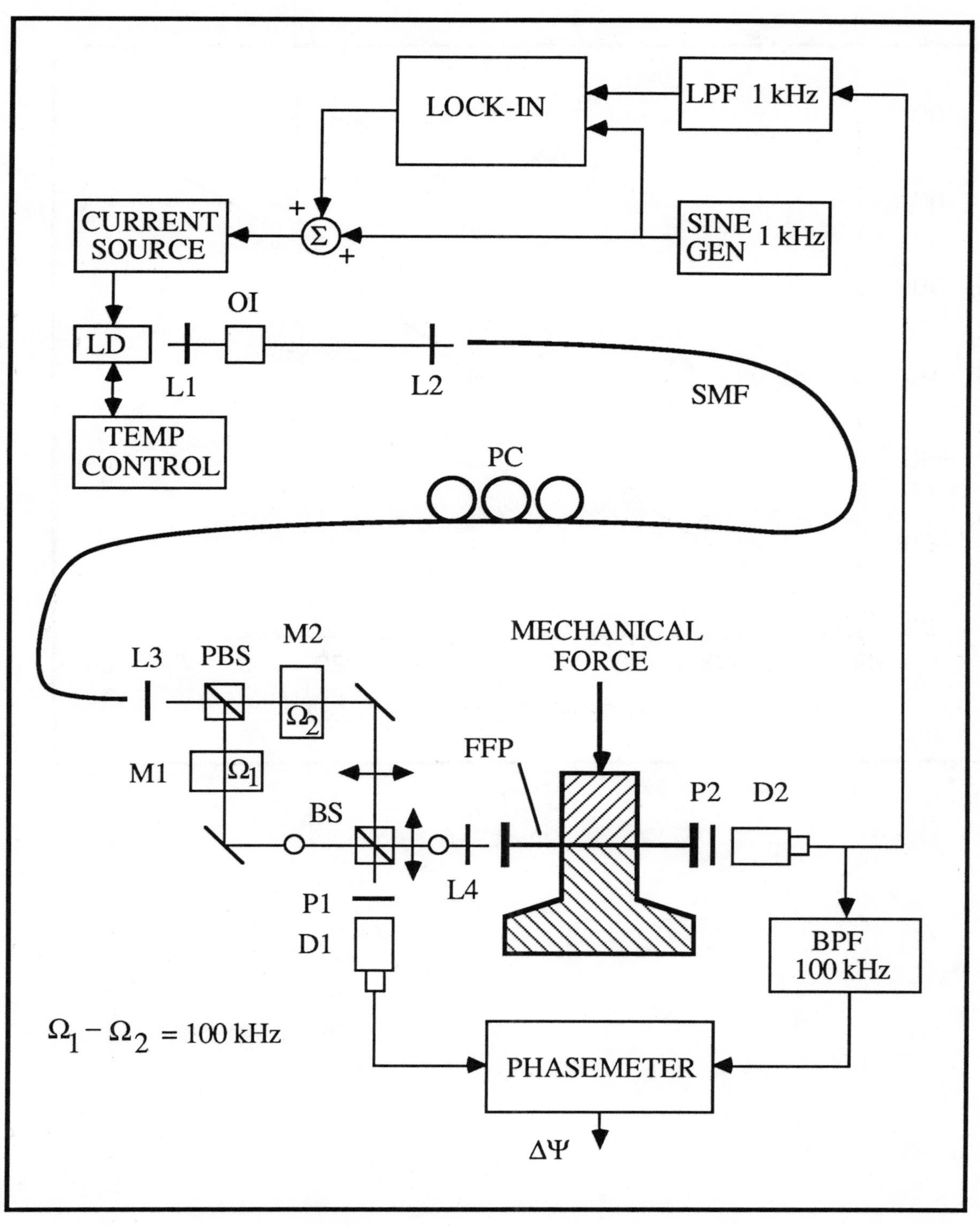
LOCK-IN
LPF 1 kHz
CURRENT SOURCE
+
Σ
+
SINE GEN 1 kHz
OI
LD
L1
L2
SMF
TEMP CONTROL
PC
MECHANICAL FORCE
M2
L3
PBS
Ω2
FFP
M1
Ω1
P2
D2
BS
L4
P1
D1
BPF 100 kHz
Ω1 − Ω2 = 100 kHz
PHASEMETER
ΔΨ

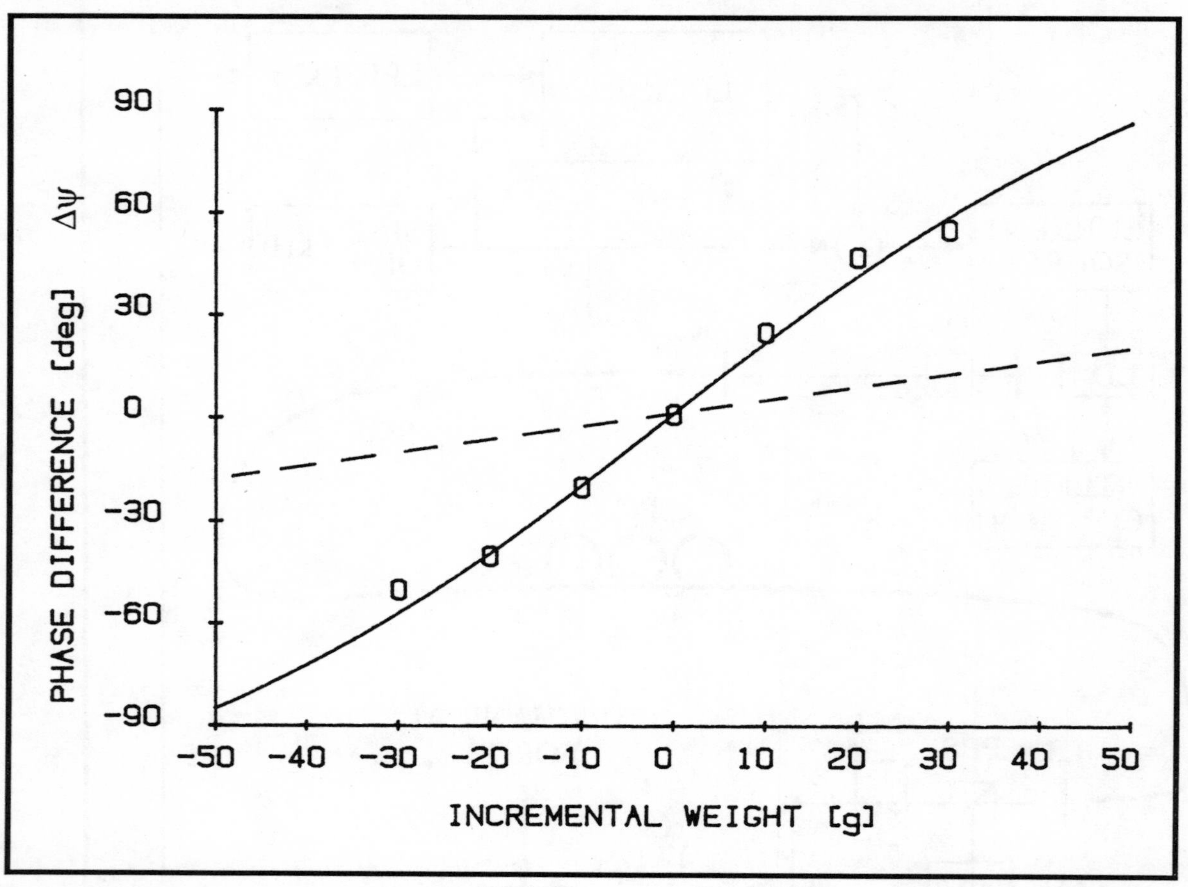

90
60
30
0
-30
-60
-90
PHASE DIFFERENCE [deg] Δψ
-50 -40 -30 -20 -10 0 10 20 30 40 50
INCREMENTAL WEIGHT [g]

OPTICALLY EXCITED MICROMECHANICAL RESONATOR PRESSURE SENSOR

K E B Thornton, D Uttamchandani and B Culshaw
Department of Electronic and Electrical Engineering
University of Strathclyde
204 George Street
Glasgow G1 1XW
Scotland

INTRODUCTION

There has been considerable interest shown in a new class of optical sensors based on the phenomenon whereby light from an intensity modulated optical source induces transverse vibrations of a micromechanical resonator on which it is incident(1). Such micromechanical resonators have most commonly been fabricated in single crystal silicon using processing steps which include photolithography, impurity diffusion and anisotropic chemical etching(2). A number of research groups have reported the optical excitation and interrogation of transverse vibrations in miniature silicon resonator structures such as cantilevers(3), diaphragms(4) and beams of complex shape(5). In all cases no more than a milliwatt or so of mean optical power was required to initiate oscillations. This power is supplied from a laser or LED, intensity modulated at the natural frequency of the resonator. The detection of the vibrations are performed interferometrically using a separate optical source and detector. For use in sensing applications the measurand must somehow create a stress in the resonator thus modifying its natural frequency. A resonant sensor of this kind which is excited by light will be electrically passive and provide an output which is a frequency. The main advantages of having frequency as the output data are easy compatability with digital processing systems, high precision measurement capability and transmission over long distances in optical fibre cables without introduction of errors due to fibre attenuation or other intensity fluctuations.

There are virtually no publications describing the use of silicon resonators excited by a few milliwatts acting as sensors. In this paper we report the first experimental results on the pressure dependence of resonant frequency of an optically excited and interrogated silicon beam resonator.

EXPERIMENT

A micromechanical silicon resonator, similar to that illustrated in Figure 1, was fabricated by silicon processing steps described in (2). The resonator is a beam clamped at both ends and spanning a pit of sides 100μm and depth of approximately 14μm. The beam has typical dimensions of 130μm x 5½μm x 2½μm and is fabricated on a silicon wafer of 65μm thickness. To operate as a pressure sensor, the beam must experience an axial force and this is achieved by mounting the wafer with the beam onto one end of a capillary tube, as illustrated in Figure 2. By applying pressure to the capillary tube, the wafer assumes a convex surface, creating a tensile stress on the outer surface where the beam is situated. By the same token a reduction in pressure creates a compressive stress in the beam.

The apparatus of Figure 3 is required for the optical excitation and optical interrogation of the silicon beam resonator. The laser diode emitting at a wavelength of 780nm is intensity modulated. An average power of 0.75mW with modulation index 0.80 is incident on the beam. A bulk optical He-Ne laser heterodyne interferometer is used for interrogation of the vibrating beam. An equivalent all fibre network is currently under assembly. By scanning the modulation frequency of the diode laser the fundamental resonance of the unstressed beam was located at 1380 kHz. The fundamental mode of vibration has a Q factor of 300 and a deflection amplitude of 50nm.

The resonant frequency of the beam is measured at values of pressure both above and below atmospheric and this data is plotted in Figure 4. The resonant frequency changes linearly with applied pressure and the frequency dependence on pressure for this resonator is 0.26kHz/mbar or alternatively 0.02%/mbar.

THEORY

The resonant frequency of a prismatic bar of rectangular cross-section subject to an axial tensile force F is determined by the equation[6]

$$f = \frac{a_o}{2\pi}\frac{t}{L^2}\left(\frac{E}{\rho}\right)^{\frac{1}{2}}\left[1 + a_1\frac{L^2F}{Ebt^3}\right] \qquad (1)$$

where a_o and a_1 are constants with values 6.45 and 0.147 respectively, L, b and t are respectively the length, breadth and thickness of the beam and E and ρ the Youngs modulus and density of the material. For a compressive force the plus sign changes to a minus sign. The beam is stressed by pressurising its wafer support. The stress on the top surface of the wafer is given by[7]:

$$\sigma = \frac{3(1 + \nu)Pr^2}{8h^2} \qquad (2)$$

where ν is the Poissons ratio of the material, r the radius of the wafer and h its thickness. The stress can be converted into a force by taking its product with the cross-sectional area of the beam A where A = bt. Substituting this, and equation 2 into equation 1 yields the resonant frequency dependence of the beam on pressure. This too is plotted in Figure 4. The discrepancy between the two results is attributed to the uncertainty in the physical constants of silicon and the precise dimensions of the resonator.

CONCLUSION

We have demonstrated an opticaly excited and optically interrogated resonant beam pressure sensor capable of measuring both positive and negative gauge pressures. The pressure coefficient of resonant frequency for this device is measured to be 0.02%/mbar. With suitable miniaturisation, sensors of this kind can be attached to the end of an optical fibre with which they are dimensionally compatible. This could lead to a new class of low cost extrinsic optical fibre sensors providing high precision measurments based on a frequency readout.

REFERENCES

1. Venkatesh S and Culshaw B, "Optically activated vibrations in a micromachined silica structure", Elect.Lett., 1985, 21, pp315-317

2. Petersen K E, "Silicon as a mechanical material", Proc IEEE, 1982 70, pp420-457

3. Hoefflin H, Kist R, Ramakrishnan S, Woelfelschneider H, Benecke W, Csepregi L, Heuberger A and Seidel H, "Optically excited micromechanical silicon vibration sensor" Proc. 4th Int. Symp on Optical and Optoelectronic Science and Eng, The Hague, March 1987.

4. Thornton K E B, Uttamchandani D and Culshaw B, "Temperature dependence of resonant frequency in optically excited diaphragms" Elect. Lett, 1986, 22, pp1232-1234

5. Andres M V, Foulds K W H and Tudor M J, "Optical activation of a silicon vibrating sensor", Elect. Lett., 1986, 22, pp1097-1099

6. Albert W C. "Vibrating quartz crystal accelerometer" Proc 28th Int. Instrum. Symp., Las Vegas, USA, May 1982, pp33-44

7. Timoshenko S P and Woinowsky-Krieger S, "Theory of plates and shells", McGraw-Hill, 1970

Figure 1. Micromechanical silicon resonator

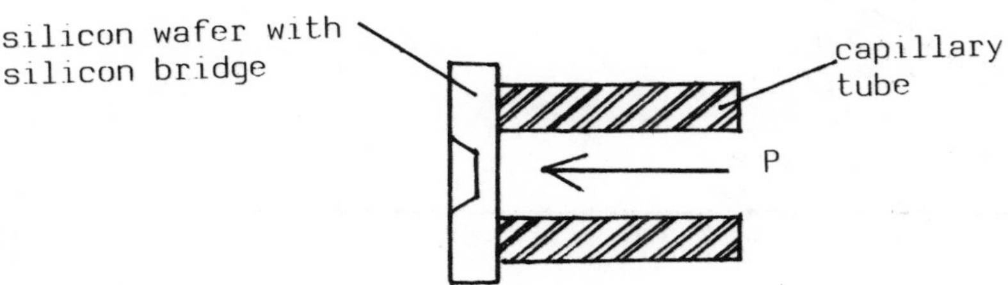

Figure 2. Silicon wafer mounted on capillary tube

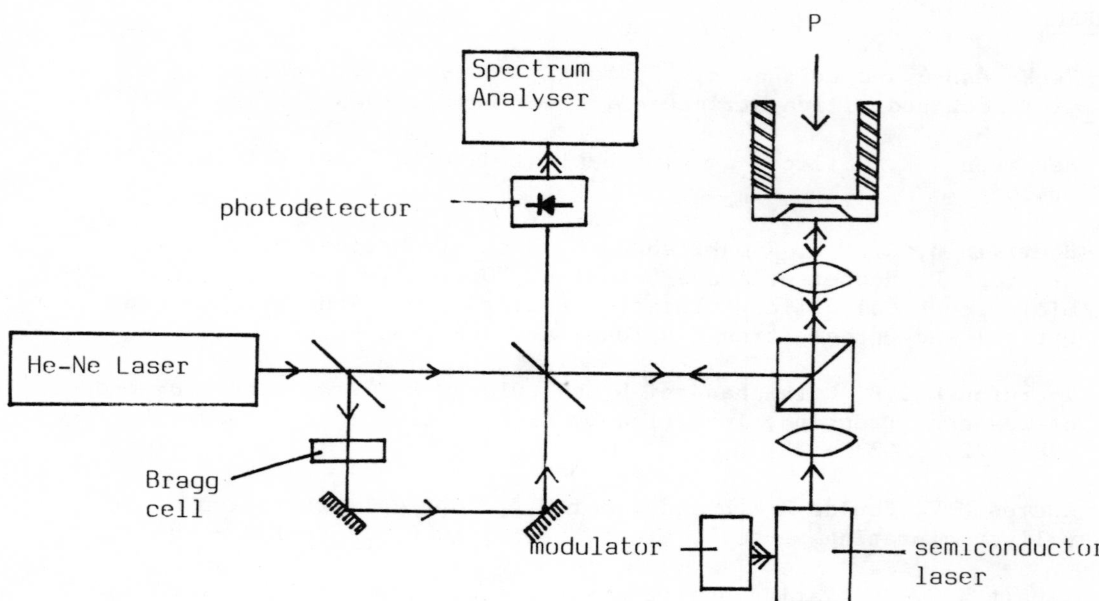

Figure 3. Bulk optic excitation and interrogation system

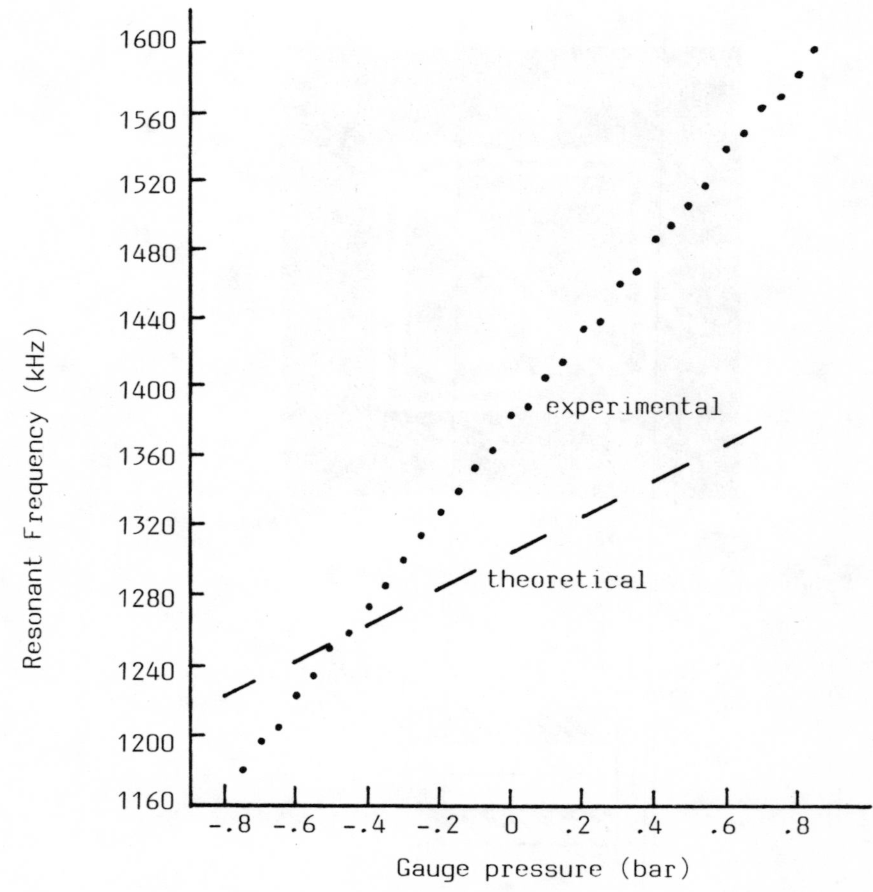

Figure 4. Resonant frequency v. Pressure

"Fiberoptic Flow Sensor"

Michael H. Ikeda, Mei H. Sun, Luxtron Corporation; and Stephen R. Phillips, Alamo Instruments

INTRODUCTION

The use of heat transfer for measuring fluid flow has a long history. Many publications in the literature cover the theory, techniques and applications of these electrically-heated anemometers in moving gases and liquids.[1,2]

Fiberoptic flow sensors are desirable when the flow rate has to be measured in a high electromagnetic field, electrical noise sensitive environments, or chemically corrosive environments. It is also desired in medical applications where electrical isolation is important for safety reasons. In this report, a small and easy to fabricate fiberoptic flow sensor is presented along with the instrumentation required for the measurement of flow rate. Catheter-based medical sensor for blood flow measurement is emphasized.

THE TECHNOLOGY

The tip of an optical thermometer is heated with near infrared radiation transmitted through the same optical fiber that optically communicates with the temperature sensor.[3] The moving fluid in its surrounding carries away heat from the sensor and causes its temperature to change. This change in temperature is directly or indirectly related to the rate of the flow.

The probe consists of a single strand of multi-mode optical fiber with temperature sensitive phosphor and infrared absorber attached to the end of the fiber tip. A short pulse of excitation light pumped into the fiber causes the phosphor to fluoresce. After the light pulse is terminated the exponential decay time of the fluorescent emission is measured, which is a function of temperature.[4] As the temperature rises the decay time shortens. The optical heating energy is provided by a continuous wave laser diode. The light from the laser impinges on the radiation absorbing material added to the sensor. The heated absorber transfers the energy to the phosphor in its vicinity. Figure 1 illustrates the schematic diagram of the optical setup.

Two basic flow monitoring techniques can be used. In the first technique, the laser diode delivers a constant amount of optical heating energy to the temperature sensing tip, and as the moving fluid carries away the heat at the sensor the temperature is lowered. The lowered temperature can be directly correlated to the flow rate. The second technique involves keeping a constant temperature at the sensor tip by modulating the duty cycle of the laser diode. The duty cycle will increase with increasing fluid velocity. For medical applications the second technique is preferred, since the controlled sensor tip temperature will prevent overheating and therefore damage of the body fluid.

EXPERIMENTAL

The first prototype, as shown in Figure 2, consists of a length of optical fiber (10 mil diameter), and a small amount of temperature sensitive phosphor (~2-3 mil diameter) attached to the center of a hemispherical lens, which is in turn attached to the end of the fiber. Within the powdered sensor and binder mixture, an infrared absorbing material is uniformly dispersed. Since the absorber also absorbs in the visible range, the amount of absorber is optimized so that adequate signal level from the phosphor is still obtained while required heating at the probe tip is achieved. Over the phosphor sensor, a thin layer of reflective coating is added to increase the sensor signal and keep the signal constant. The fiberoptic sensor is then threaded into a Swan-Ganz catheter and the sensor tip potted in place. A section of the tip is sitting near the outside surface of the catheter and exposed to the flowing fluid to provide good heat transfer.

The assembled catheter sensor is tested in the laboratory in a flow loop where 0 - 100 cm/sec (or 0 - 20 l/min) flow rate can be controlled. The instrument employed is a modified Luxtron Fluoroptic thermometer Model 750. The correlation curve of the laser diode duty cycle with flow rate of water using a 40 mw, 820 nm laser diode is shown in Figure 3. With 1 second integration time, a resolution of 1 cm/sec is achieved.

REFERENCES

1. P. Freymuth, "Feedback Control Theory for Constant-Temperature Hot-Wire Anemometers", _The Review of Scientific Instruments_, Vol. 38, No. 5, 1967, pp. 677-681.

2. W.A. Seed and N.B. Wood, "Development and Evaluation of a Hot-Film Velocity Probe for Cardiovascular Studies", _Cardiovascular Research_, Vol. 4, 1970, pp. 253-263.

3. Stephen Phillips, U.S. No. 4,621,929, Nov. 11, 1986.

4. K.A. Wickersheim, S.O. Heinemann, H.N. Tran, and M.H. Sun, "A Second Generation Fluoroptic Thermometer", ISA 1985 paper no. 85-0072.

Michael H. Ikeda
Fiberoptic Flow Sensor

CAPTIONS

Figure 1: Modified Model 750 Head Utilizing Decay time Technology and Infrared Laser Diode Heating.

Figure 2: Fiberoptic Catheter Flow Sensor.

Figure 3: Laser Duty Cycle versus Flow Velocity; Catheter Sensor in Flowing Water.

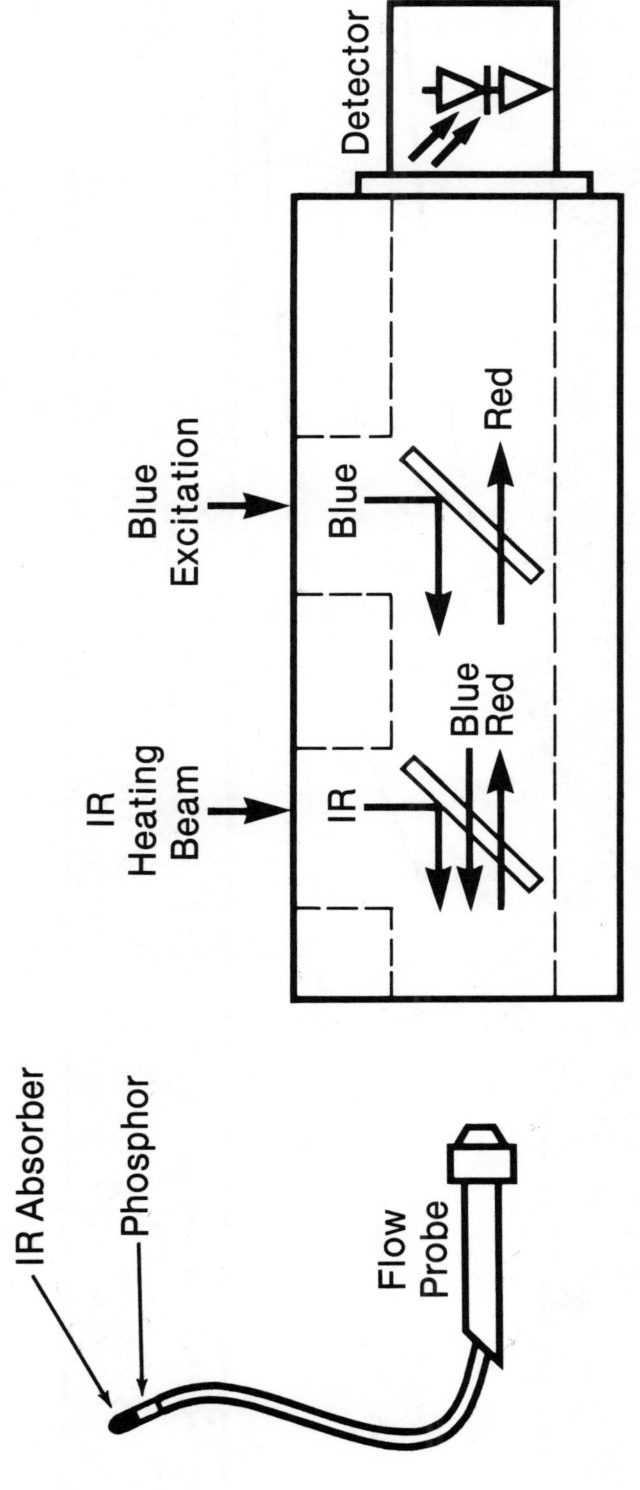

Figure 1. Modified Model 750 Head Utilizing Decay Time Technology and Infrared Laser Diode Heating

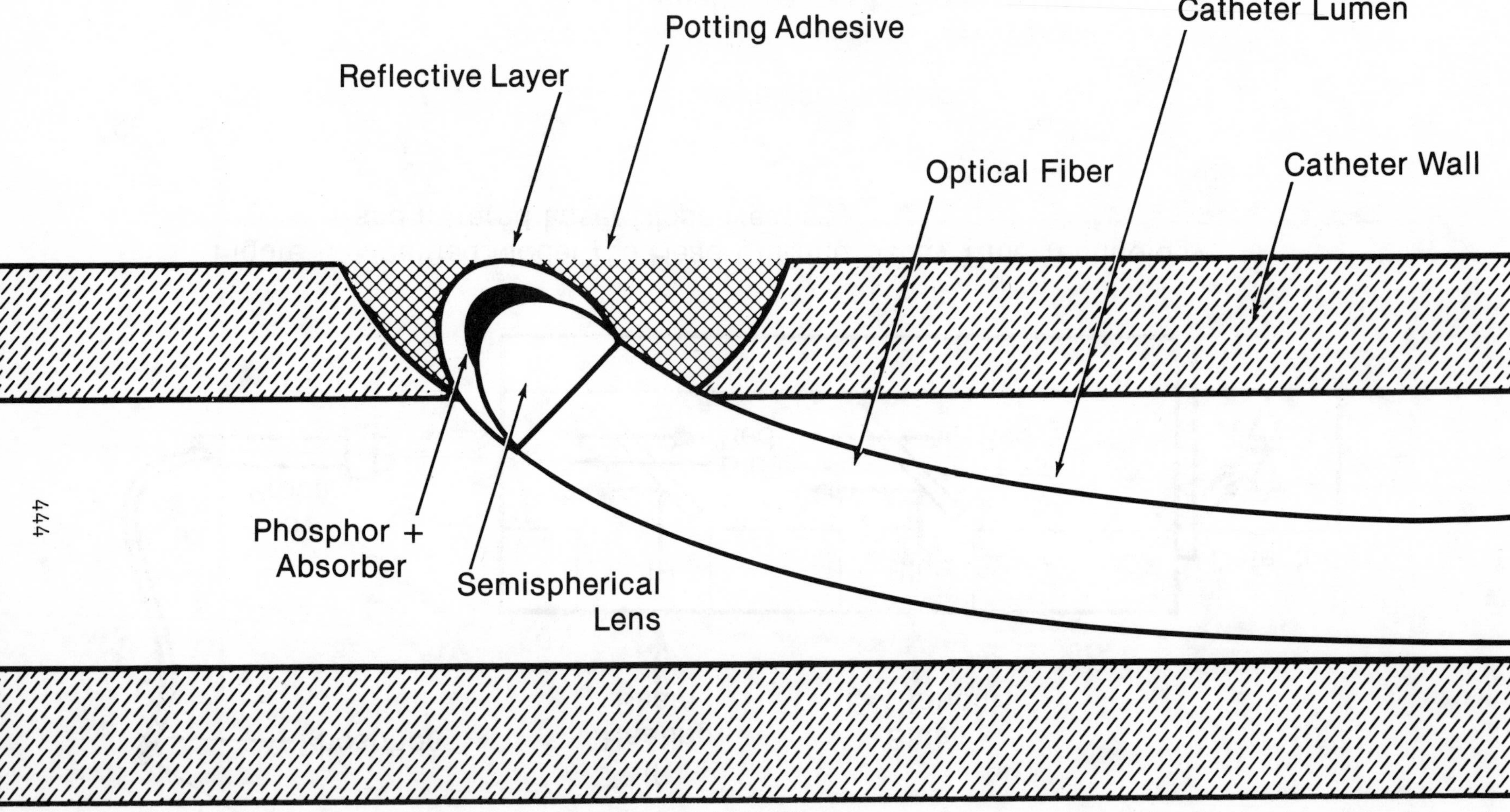

Figure 2. Fiber Optic Catheter Flow Sensor

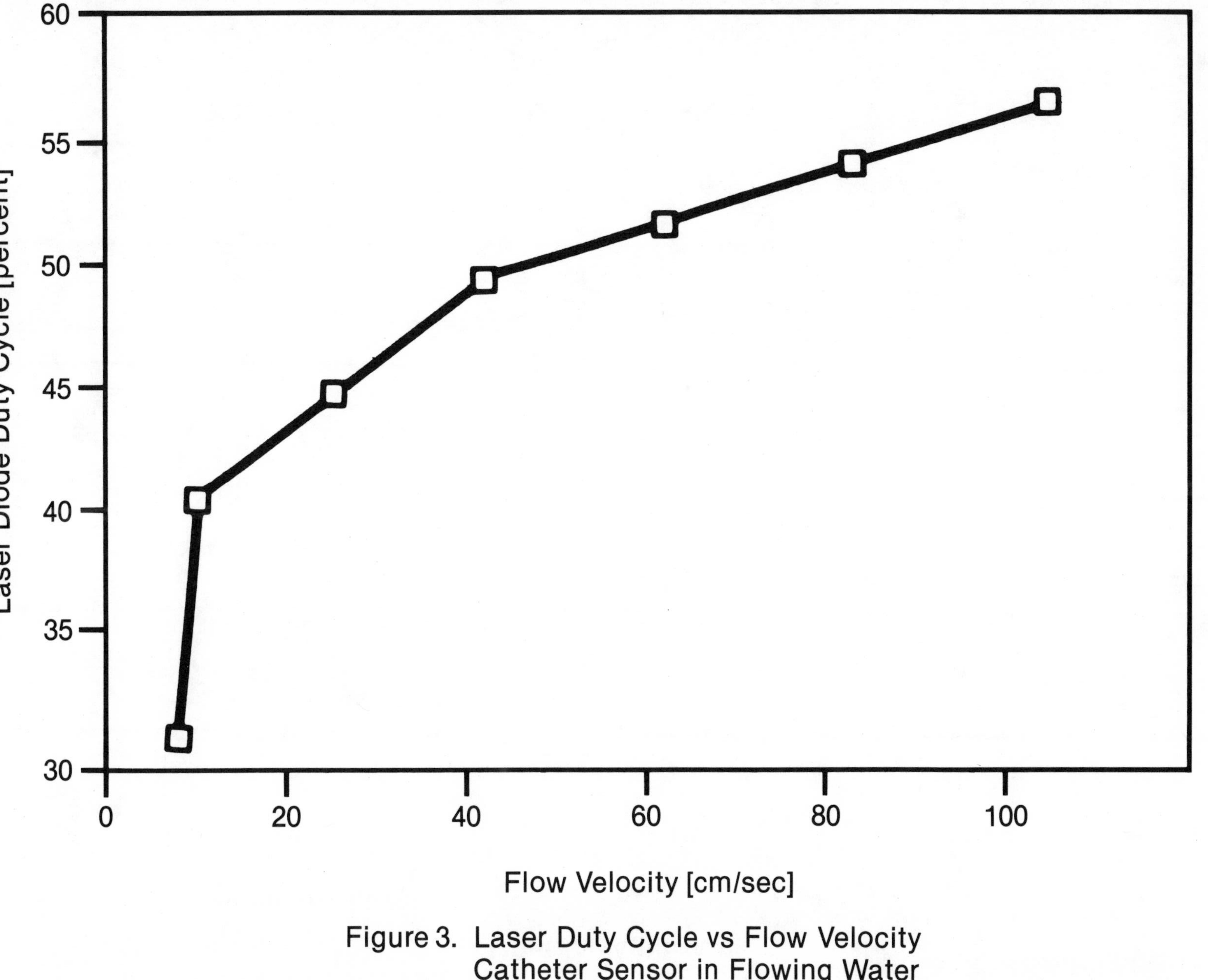

Figure 3. Laser Duty Cycle vs Flow Velocity Catheter Sensor in Flowing Water

FRIDAY, JANUARY 29, 1988

MEETING ROOM 5-7-9

10:30 AM–12:00 M

FDD1–5

SOURCES FOR SENSORS

Scott Rashleigh, Australian Optical Fiber Research Party Ltd., Australia, *Presider*

E. Snitzer, Rare Earth Fibers

The earlier work on soft glass fiber lasers doped with rare earths were used as oscillators[1] and for several amplifier applications. A gain of 47dB was obtained for the output at 1.06μm from an $InAs_{1-x}P_x$ laser diode.[2] A gain of 40dB was obtained for a helium neon laser emitting at 1.0621μm.[3] Experiments were done with the use of a single mode fiber as a preamplifier in a detection system.[4] Spontaneous emission and saturation limits imposed by super luminescence have also been considered.[5] With the availability of high radiance lasers and the use of low loss glass hosts end pumped Nd doped fibers were lased first in multimode fibers[6] and later in single mode fibers with Nd, Pr or Er.[7,8]

In the neodymium and ytterbium fibers discussed here the predominent constituent of the glass is SiO_2, but small quantities of other ions such as GeO_2, P_2O_5 or Al_2O can be added with the glass still compatible with SiO_2 cladding.[9] The small additions can increase the index of refraction for the core, effect the softening point, increase the rare earth solubility limit, and shift the peak of the emission wavelength for neodymium in the $^4F_{3/2}$-$^4I_{11/2}$ transition. Depending on the constituent that is added, and the pump wavelength, the peak wavelength for the emission can be varied from 1.054μm to 1.114μm. The largest concentration of rare earth was made possible by the addition of aluminum. Concentrations in excess of 14 wt % of Yb_2O_3 have been added by MCVD to Al-Si glass. The presence of aluminum broadens the 1.06μm fluorescence substantially and furthermore gives a strong dependence of the peak fluorescence wavelength on the pump wavelength when pumping into the band in the vacinity of 0.8μm. Shifting the pump wavelength from 0.806 to 0.846μm causes the peak fluorescence to shift from 1.064 to 1.114μm.

Neodymium fibers have been lased at .905, 1.06, and 1.4μm and have been operated as super luminescent sources and amplifiers, as well as oscillators.[10] The measured gain per absorbed pump power at 0.8μm was 0.47 dB/mw for a fiber with an NA of 0.15 and a V value of 1.75 at 1.06μm. Experiments with lasers containing ytterbium or co-doped with neodymium and ytterbium will be described.

E. Snitzer, Rare Earth Fibers

1. C.J. Koester and E. Snitzer, Appl. Opt. 3, 1182 (1964).

2. B. Ross and E. Snitzer, IEEE J. Quantum Electron. QE-6 361 (1970).

3. G.C. Holst, E. Snitzer, and R. Wallace, IEEE J. Quantum Electron. QE-5, 342 (1969).

4. G.C. Holst and E. Snitzer, IEEE J. Quantum Electron. QE-5, 319 (1969).

5. E. Snitzer, Proc. of First European Electro-optics Markets and Technology Conf., IPC Science and Technology Press, Geneva (1972).

6. J. Stone and C.A. Burrus, Appl. Phys. Lett. 23, 388 (1973).

7. R.J. Mears, L. Reekie, S.B. Poole, and D.N. Payne, Electron. Lett. 21, 738 (1985).

8. S.B. Poole et al. J. Lightwave Technology. LT-4, 870 (1986).

9. See Also C.A. Millar, B.J. Ainslie, I.D. Miller, S.P. Craig, Paper W14, OFC/IOOC '87.

10. H. Po, F. Hakimi, R.J. Mansfield, B.C. McCollum, R.P. Tumminelli and E. Snitzer, OSA Annual Mtg. Seattle (October 1986)

High Power GaAlAs Superluminescent Diode For Fiber Sensor Applications

Norman S.K. Kwong, Kam Y. Lau, Nadav Bar-Chaim,
Israel Ury and Kevin Lee
Ortel Corporation, 2015 West Chestnut Street
Alhambra, California 91803-1542
818-281-3636

Superluminescent Diodes (SLD's) are the key element in certain fiber optic sensors such as fiber gyroscopes [1-3]. The broad-band characteristics of SLD's reduce the noise due to Rayleigh backscattering, polarization noise, and bias offset due to the nonlinear Kerr effect in fiber optic sensor system. Therefore, there is a need for a high power, high efficiency SLD which emits into a stable, single spatial mode for such applications. The optimal approach to design a SLD is to base on a proven high power, high efficiency "window" type buried heterostructure laser which is capable of emitting greater than 50 mW CW into a stable index-guided lateral mode. Lasing is suppressed for SLD operation by antireflection (AR) coating the emitting facet and by incorporating a rear absorber section. The resulting device emits high optical power 22.5 mW in the SLD mode at very low injection current 100 mA. The spectral modulation depth is below 25% over the entire emission spectral bandwidth of 20nm, with a symmetrical beam divergence (20°x40°) and a stable transverse mode.

The structure of the SLD is shown in Fig. 1. It is based on the "window" buried heterostructure laser [4]. The buried heterostructure provides excellent current confinment, and therefore, low operating current. This real index guiding laser structure also provides for stable single transverse mode output over a large range of operating current. The output region near the front facet is composed of a layer of unpumped GaAlAs which forms a transparent window thus averting catastrophic damage suffered by conventional GaAs/GaAlAs lasers at high power. An anti-reflection coating on the window facet minimizes feedback to the active region and therefore enhances optical output power from the front facet. An absorbing region at the back end of the active gain region absorbs radiation propagating away from the output end, thereby suppressing laser oscillation. The absorbing region is formed by shorting the p-n junction to ground in the rear section of the waveguide. This is an efficient method to increase the absorption of the waveguide by "draining" the photo-excited carriers to ground. Typical device dimensions are: length of the pumped region L=250 μm ; the grounded absorber region l=200 μm ; the gap between the pumped and the grounded absorber region d=50 μm ; and the window region, t=5 μm .

The light-current characteristics of the device is shown in Fig. 2. The device is capable of putting out 22.5 mW at 100 mA injection current. The pulse light-current curve shows that the device can operate in the SLD mode at least up to 30 mW. This implies that the CW output power of the device is limited by heating, since it is mounted junction side-up on a copper heat sink. With junction side down mounting, we expect the output power can be futher improved. The emission spectrum at 22.5 mW is shown in Fig. 3. The spectral width at FWHM is 20 nm. The estimated coherence length is about 35 μm. The spectral modulation, defined as $m=(I_{max}-I_{min})/(I_{max}+I_{min})$ at the peak of the spectrum, is less than 25%.

The far-field pattern of the SLD is shown in Fig. 4. With FWHM of 20° and 40° for parallel and perpendicular scans respectively, the smooth angular dependence is an indication of the very low coherence and its stability with increasing current indicates the presence of real index guiding. The degree of polarization varies from 45% to 70% for the current range from 10 mA to 100 mA.

The stable fundamental transverse mode behavor of the SLD allows for efficient coupling to optical fibers. We have obtained 20% coupling efficiency with the SLD butt-coupled to a 0.23NA, 50 μm diameter core graded index multimode fiber. For coupling light into single mode polarization preserving fiber, we have used a combination etching and arc-melting technique to prepare microlens-tipped fibers. We have obtained 20% coupling efficiency into a .11NA, 4.5 μm diameter core fiber (Hitachi SPPF).

This research is supported by U.S. Naval Research Laboratory under Contract N00014-86-C-2477.

Reference:

1. K. Bohm, P. Marten, K. Petermann, E. Weidel, and R. Ulrich, Electron. Lett. **17**, 352, 1981.

2. W.K. Burns, C.L. Chen, and R.P. Moeller, J. Lightwave Tech. **LT-1**, 98, 1983.

3. S. Tai, K. Kojima, S. Noda, K. Kyuma, K. Hamanaka, and T. Nakayama, Electron. Lett. **22**, 546, 1986.

4. K.Y. Lau, N. Bar-Chaim, I. Ury, and A. Yariv, Appl. Phys. Lett. **45**, 316, 1984.

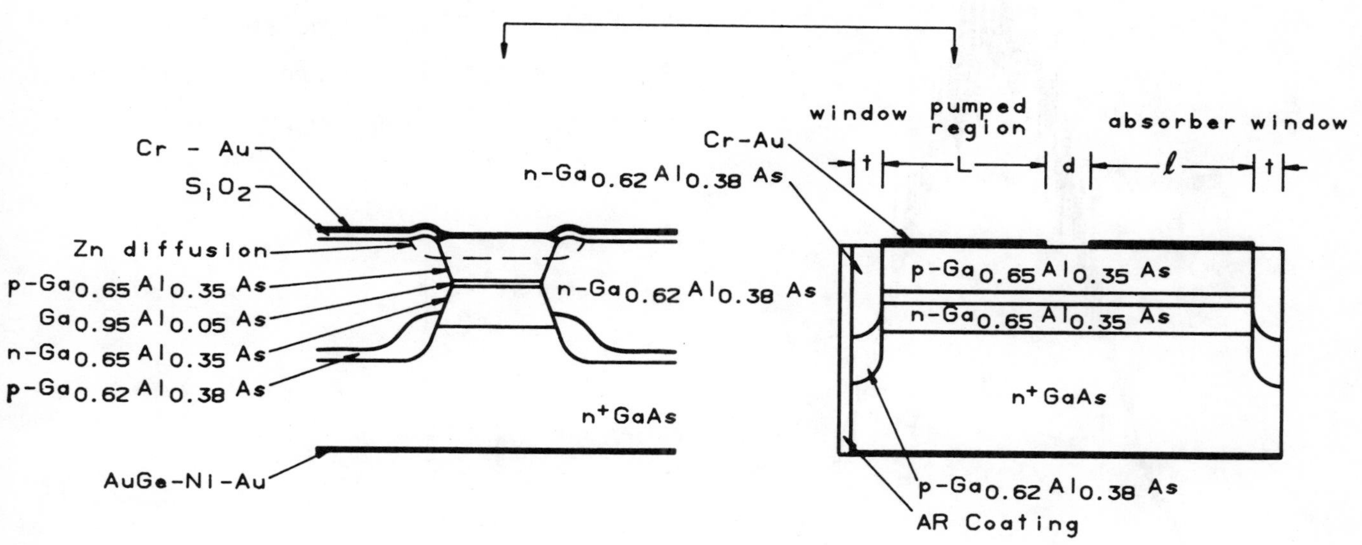

Fig. 1: Schematic diagram of the window buried heterostructure SLD.

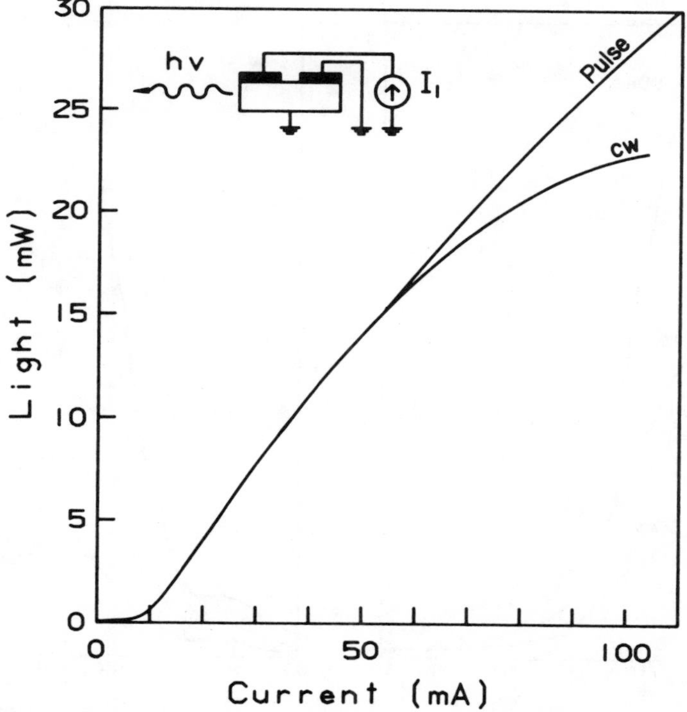

Fig. 2: Light versus current characteristics of the SLD for both CW and pulse operation.

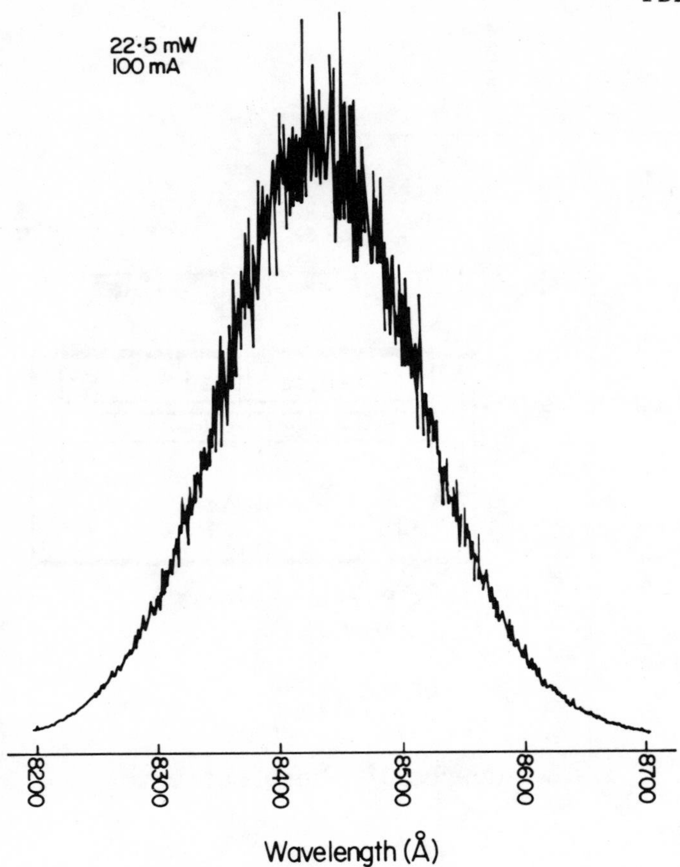

Fig. 3: The spectrum of the SLD at current 100 mA and 22.5 mW output power.

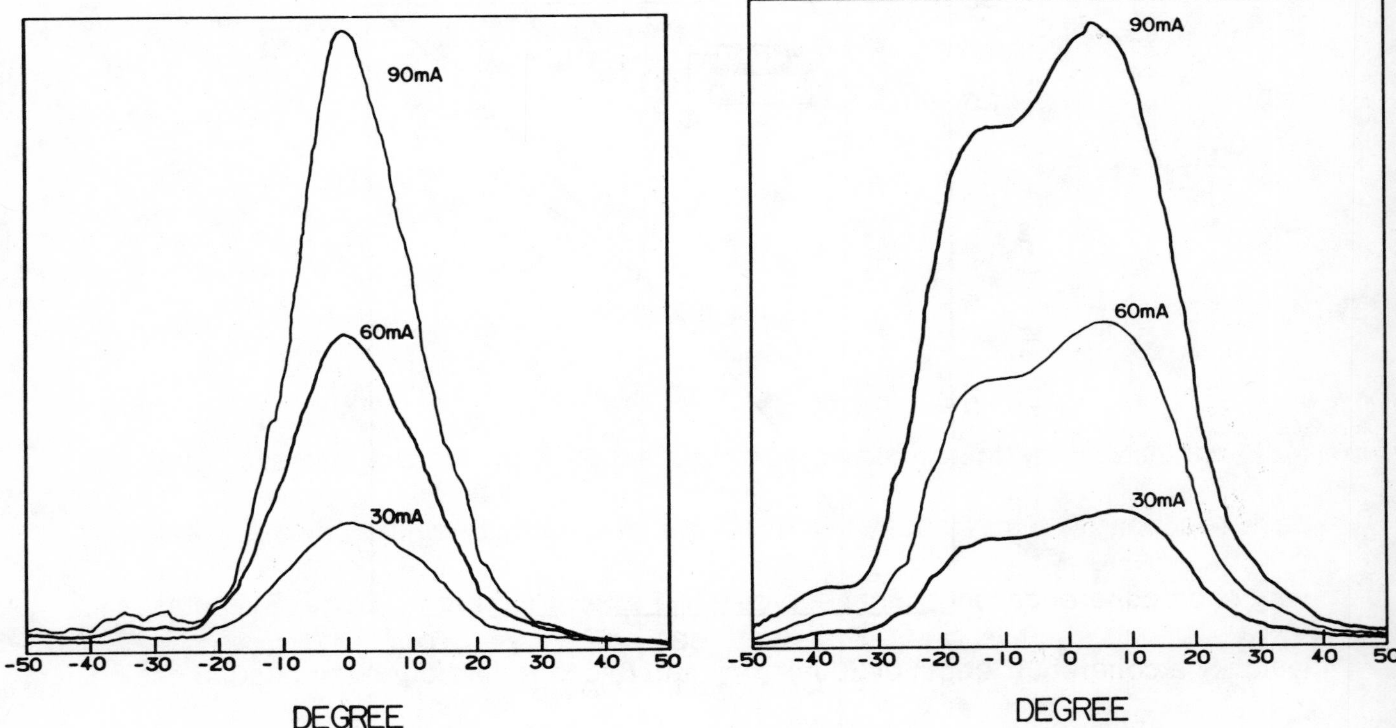

Fig. 4: Far-field intensity distribution of the SLD: (a) parallel to the junction; (b) perpendicular to the junction.

A CHIRPED SEMICONDUCTOR LASER AS AN ALTERNATIVE TO THE SLD IN A FIBER GYRO

L. Hergenroeder, S.P. Smith, and S. Ezekiel

Research Laboratory of Electronics,

Massachusetts Institute of Technology

Cambridge, MA 02139

(617) 253-3783

ABSTRACT

A chirped semiconductor laser is used instead of a superluminescent diode (SLD) to reduce the effects of backscattering and the optical Kerr effect in a fiber interferometer gyroscope.

SUMMARY

In a multiturn fiberoptic interferometer gyroscope, the effects of backscattering, backreflections and the optical Kerr effect are all virtually eliminated by the use of a very short coherence length light source. The superluminescent diode, SLD, has typically a coherence length of 30 microns and has therefore been the popular choice among researchers. The SLD, however, has some disadvantages, namely, a short lifetime because it operates way over threshold, a low power output, a low coupling efficiency into a single mode fiber, and is rather expensive since the demand for SLDs is limited.

As an alternative to the SLD, we have investigated the use of a frequency modulated, or chirped, semiconductor laser. Our experimental fiber gyro set up was a simple one. A 730 m long non-polarization preserving fiber was wound on a 30 cm diameter drum. The primary coupler was a 3 dB all fiber coupler which is coupled into the fiber coil by means of mechanical splices. A conventional fiber polarization controller was used to adjust the polarization within the loop and the phase modulator was a PZT cylinder around which a short length of fiber was wound. A bulk optic polarizer with an extinction ratio of 10^{-6} was placed before the input port of the coupler.

The laser diode was a Hitachi HLP-1400 operating at a single frequency around 830 nm. The laser spectral width was broadened up to 35 GHz by modulating the laser current at a rate 20 kHz. With such a source, we were able to achieve short term noise of 0.13 deg/hour for an averaging time of 0.1 sec, which was close to the shot-noise-limit. This translates to 7×10^{-4} deg/root hour.

The effect of chirping the laser on the optical Kerr effect was studied by inducing a difference in intensity in the counter propagating beams, measuring the resulting nonreciprocal offset and then reducing this offset by varying the laser spectral width through chirping. In our set up, the optical Kerr effect was about 1 deg/hour for a 1 μwatt power difference. Starting with a Kerr effect offset of 36 deg/hour, we measured the reduction in this offset as a function of the laser linewidth. We found that a broadening of 35 GHz reduced the offset to a negligible value. The laser spectral width was measured with a Fabry-Perot interferometer.

These preliminary experiments show that a simple chirped semiconductor laser could be a good alternative to the SLD in reducing backscattering and the optical Kerr effect. In addition, the advantages of higher power and greater coupling efficiency into a single mode fiber together with lower cost and longer life makes the chirped laser even more attractive for gyro applications.

This research was supported by the Joint Services Electronics Program.

OPTICAL FEEDBACK EFFECTS ON SUPERLUMINESCENT DIODES

R.O. Miles*, W.K. Burns, R.P. Moeller and A. Dandridge

Optical Techniques Branch
Code 6570
Naval Research Laboratory
Washington, D.C. 20375
Tel (202) 767-3298
* Sachs/Freeman Associates

INTRODUCTION:

The effects of optical feedback on broadband sources such as the Superluminescent Diode, SLD, have not been previously reported. With rapid development of the fiber optic gyroscope where these devices are used as sources these effects may tend to limit the performance of such systems.

There are two sources of optical feedback in fiber optic gyroscopes. Rayleigh scattering, which is a distributed effect through out the fiber length and direct feedback which results from the loop configuration of the gyro. Rayleigh scattering appears as a temporal fluctuation that varies with time. This behavior is due to the environmental effects on the scattering centers in the fiber, such as thermal and acoustic perturbations which effect the mechanical structure of the fiber on a microscopic scale. Direct feedback occurs due to the non-reciprocal nature of light propagating in the fiber loop during rotation of the gyro. This effect is seen as a voltage dependent on the rotation rate which induces an error in the measured signal[1]. In both of these cases light fed-back into the gyro source induce spectral changes to occur within the device as well, which would also affect the performance of the system.

EXPERIMENTAL PROCEDURE:

The experimental arrangement used to measure the effect of Rayleigh scattering on the SLD is shown in fig. 1. A General-Optronics SLD with fiber pigtail operating at 830 nm was used as a source in this experiment. The fiber used was Corning single-mode, polarization maintaining fiber at 830 nm. A 1 meter fiber pigtail was attached by General Optronics. Provision was made to observe the spectral output of the laser both through the attached fiber and after inserting a 2 Km length of the same fiber.
The output of the fiber in each case was terminated using index matching fluid to prevent reflections from the fiber end back into the SLD device.

The output of the fiber was analyzed using a Spex 3580 spectrometer whose output was digitized and then analyzed by computer. The spectral output region of the SLD was scanned on the spectrometer, acquired, amplified, digitized and displayed as a normalized intensity as a function of wavelength on the computer. A spectral analysis of the data is performed using a smoothing algorithm. Using a filtering function which filters the high frequencies an "average" spectral curve can be determined which simplifies the procedure in approximating the half-power points and peak (center) wavelength of the spectral envelope. By noting the difference between the half-power points one can infer the overall spectral width of the envelope. By noting the position of the center wavelength it can be determined if the spectral envelope has shifted due to the effects of optical feedback.

The spectral output of the General Optronics device SAR 93-1 is shown in fig. 2. This device is capable of operating at 3 mw out of the fiber with an estimated 20% coupling. In this figure we show the output running at 1 mw fiber-output at 178 ma with the fiber terminated. the closely spaced modal structure observed across the spectra is due to the longitudinal mode spacing of the SLD cavity. The more prominent structure is due to reflections set up by the near end of the fiber. With out knowing more about the fiber coupling configuration it is difficult to estimate how much light is being coupled back into the SLD to set up this kind of structure.

When 2 Km of polarization maintaining fiber is added a spectral response seen in fig. 3 is observed. In this case index matching fluid was used to match the gap between the fibers where the fibers were butted together and at the fiber output. No attempt was made to align the fiber axes. As can be seen the modal distribution of the spectral output has been significantly altered. It appears that the Rayleigh back-scattering sets up a fine structure, which may be time varying in nature, and which overwhelms the feedback due to by the near end of the fiber. Rayleigh back-scatter has been measured in single mode fibers to be on the order of -30 dB[2]. Assuming coupling losses at less than 1 db between the two fibers and 20% at the laser, the back-scattered light coupled into the SLD would be on the order of a micro-watt.

Interestingly when the fiber axes are aligned within a few degrees, a stronger interaction is observed insofar as the spectral envelope now consists almost entirely of fine structure set up by the Rayleigh backscatter. The structure due to the fiber near-end reflections has almost disappeared, as seen in fig. 4. Since coupling between nonaligned axes of the fiber is quite high (less than 1 dB loss) the effect observed here is believed due to more favorable polarization coupling. In this case more backscattered light is being coupled to the same polarization state of the SLD and therefore inducing a stronger interaction.

CONCLUSIONS:

We have observed the effects of Rayleigh back-scattering coupled back from 2 km of fiber into a high-power SLD. We have noted that this time-varying source of feedback tends to be distributed across the entire spectrum and has a "smoothing" effect on spectral structure caused by direct feedback, in this case from the near end of the fiber. It is also important to note that the SLD device tested demonstrates a favored polarization state insofar that light coupled from fibers whose axes are aligned with the SLD have a predominate effect.

The observations of the effects of direct feedback on the SLD spectra will also be reported. Implications of these effects on the fiber gyro performance will be discussed.

REFERENCES

1. Moeller, R.P, Burns, W.K. and Frigo, N.J. Scale factor Accuracy and Stability in an Open Loop Fiber Optic Gyroscope, submitted to this conference

2. Yurek, A. and Dandridge, A. Optical and Electronic Reduction of Rayleigh Backscatter Noise in Fiber-coupled Semiconductor Diode Lasers. Electron. Lett. 1986, 22, PP.645-647

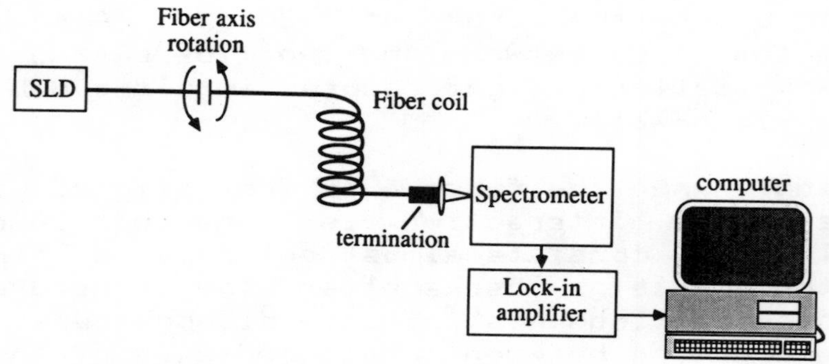

Fig. 1 Experimental configuration.

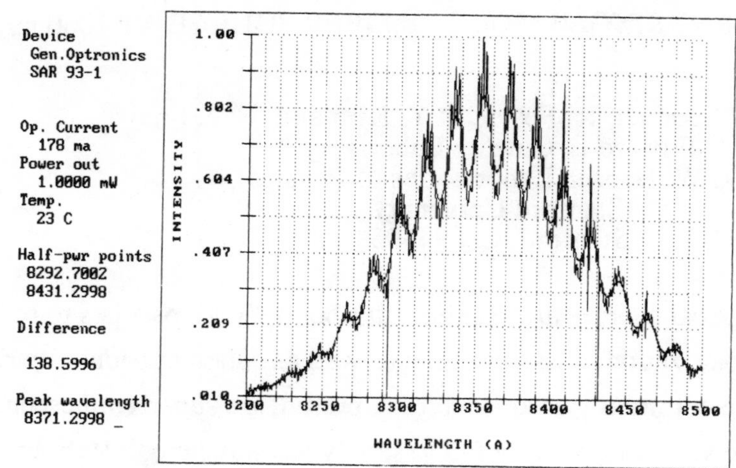

Fig. 2 Spectral output of pigtailed SLD operating at 1 mw. with output fiber terminated.

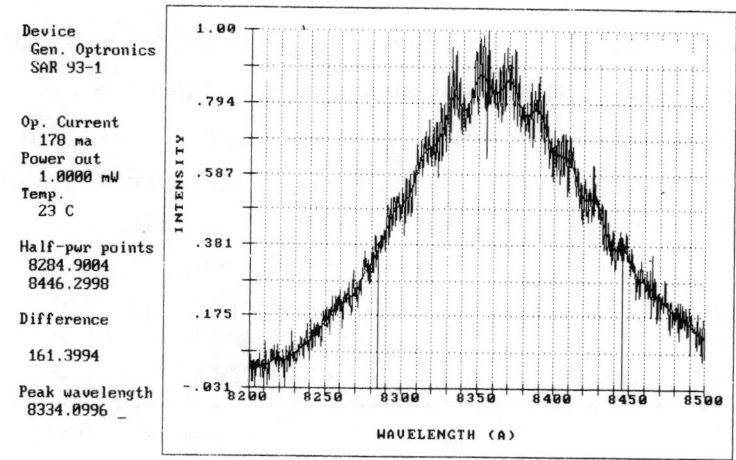

Fig. 3 Spectral output of pigtailed SLD coupled to a 2 km fiber. with output fiber terminated axes not aligned.

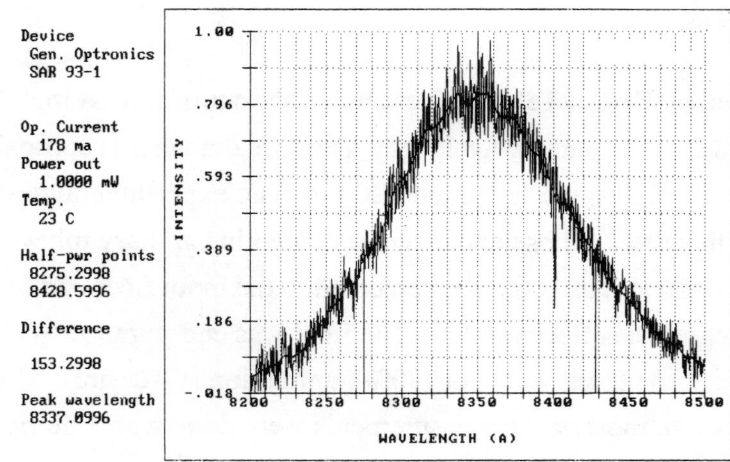

Fig. 4 Spectral output of pigtailed SLD coupled to a 2 km fiber, with output fiber terminated axes aligned.

SUPERFLUORESCENT SINGLE MODE Nd:FIBER SOURCE AT 1060 nm

K. Liu, M. Digonnet, K. Fesler, B.Y. Kim and H. J. Shaw
Edward L. Ginzton Laboratory
W. W. Hansen Laboratories of Physics
Stanford University
Stanford, CA 94305

Miniature broadband optical sources are desirable in fiber optic gyroscopes to reduce non-reciprocal phase error due to polarization cross-coupling, as well as coherent backscattering noise and the optical Kerr effect.[1] The commonly used sources at present are superluminescent laser diodes (SLD). However, these sources generally fail to satisfy the wavelength stability requirement, as their emission wavelength is very sensitive to temperature and optical feedback. In addition, they incur high coupling loss into single-mode fiber, yielding typically a few hundreds of μW of usable power.

Recently, neodymium-doped silica singe-mode fiber emerged as another potential miniature fluorescent source.[2] The potential of such a source was also investigated theoretically[3]. In principle it can provide significant output of 1060 nm emission when optically end-pumped hard enough to be in the stimulated emission regime. In this paper we report high power operation (> 10 mW) of a fiber superluminescent source and discuss its wavelength stablility.

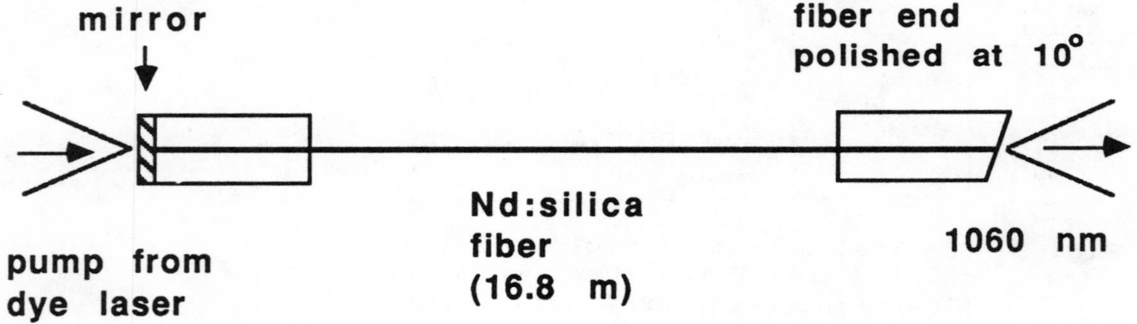

Fig. 1 Experimental device

The active fiber, provided by British Telecom Research Laboratories, was single mode at 1060 nm but supported two modes at the pump wavelength. The core diameter is approximately 6 μm and the cut-off of the second order mode is around 920 nm. The experimental device, shown in Fig. 1, consists of 16.8 m of this fiber, whose ends were mounted in capillary tubes and mechanically polished. A high reflector for 1060 nm, cemented to the input fiber end, restricts the output to one direction and increases the effective gain. The fiber was end-pumped by a cw Styryl 9 dye laser coupled into the fiber through the reflector (80% tranmission at 820 nm). Although the peak absorption was around 804 nm, most of the measurements were done with a pump wavelength around 820 nm where the dye laser output was strongest. Most of the pump power launched into the fiber was absorbed in the length used. Of the two strongest emission lines of Nd:SiO_2 (about 900, 1060 nm), only the 1060 nm line reflected by the mirror had large net gain. The 900 nm line was also strongly self-absorbed in the fiber. Weaker emission lines at longer wavelengths were not seen by the silicon detector. Therefore, most of the measured output power (>99%) consisted of the 1060 nm emission. To avoid resonant oscillation between the mirror and the output fiber end, the latter was polished at a 10° angle.

The amount of pump power coupled into the fiber was estimated by measuring the coupling efficiency of the pump beam into a short (< 10 cm) section of the same fiber, in which the pump absorption was negligible (estimated to be less than 3%). The coupling efficiency, not including the mirror reflection, was typically in the range of 50-70%. The absorbed pump power was then calculated from the measured incident and unabsorbed pump powers (the latter was in fact less than 5% of the absorbed power). Fig. 2 shows experimental data points of output power vs. estimated absorbed pump power.The fiber output was found to be essentially unpolarized, although more detailed measurements are needed to confirm this observation. At 10 mW output power, the conversion slope efficiency was 45% with a net conversion efficiency between the absorbed pump power and the 1060 nm signal of 12%.

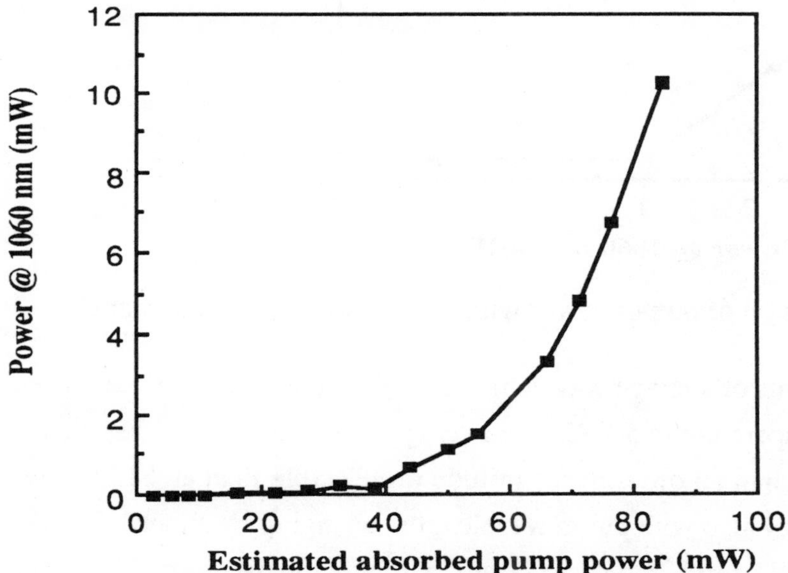

Fig. 2 Output vs input pump power

The exponential behavior of the output power in Fig. 2 can be understood as follows. For low pump power, the output power increases linearly with pump power due to additive spontaneous emission. As stimulated emission becomes significant, the output increases nonlinearly essentially as e^g. The gain g is the linecenter gain experienced by a signal which has travelled the whole length of the device. Placing the mirror at one end of the fiber effectively doubles the active length and, barring any saturation effects, doubles the gain. It should be noted that this is a very large effect as the output now increases as e^{2g}. This behavior was qualitatively observed: the output power was found to drop to about 300 μW for 80 mW absorbed pump when the mirror was removed.

As the output power increases, the linewidth narrows due to greater amplification at the peak of the output spectrum. As Fig. 3 shows, the measured linewidth narrows asymptotically and becomes nearly constant at about 16 nm for total output power in the mW range. A typical output spectrum is shown in Fig. 4.

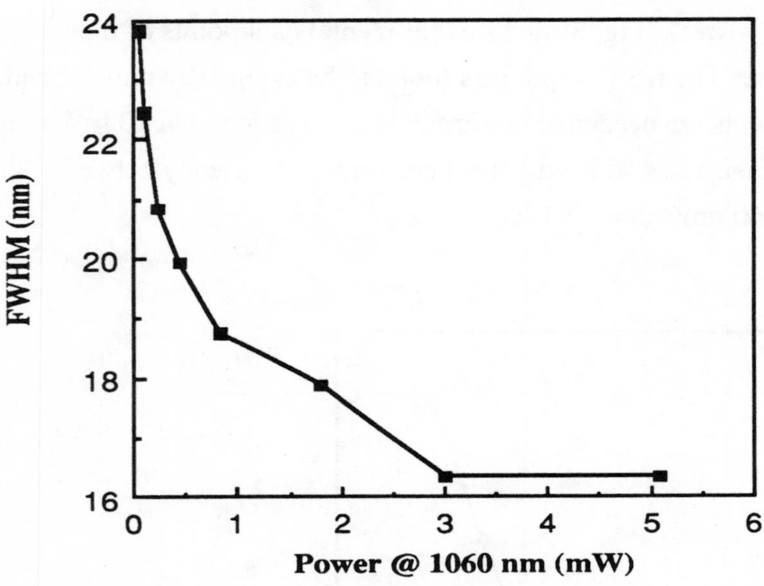

Fig. 3 FWHM as a function of output power with pump wavelength at 825 nm

Preliminary measurements of average wavelength shift with temperature were made. Upon heating the fiber from room temperature to 50° C, the average wavelength was observed to shift by a few ppm/°C. This is greater than an order of magnitude more stable than an SLD. The average wavelength was observed to vary also with pump wavelength, changing by about 1.5 nm when the pump wavelength was varied from 808 nm to 821 nm by tuning the dye laser. Therefore, the pump laser wavelength should be controlled to several tens of ppm for the wavelength stability to equal the intrinsic temperature stability. This issue is particlarly relevant in the case of laser diode pumping. More detailed characterization of the spectral stability is under way.

In summary, we present high power operation of a Nd-doped fiber superluminescent source with broad spectrum suitable for use in fiber gyroscopes. Since the source is a fiber, almost all of the optical power can be coupled into a single-mode fiber system. Preliminary measurements indicate that the spectral stablility of the source may be much better than that of an SLD.

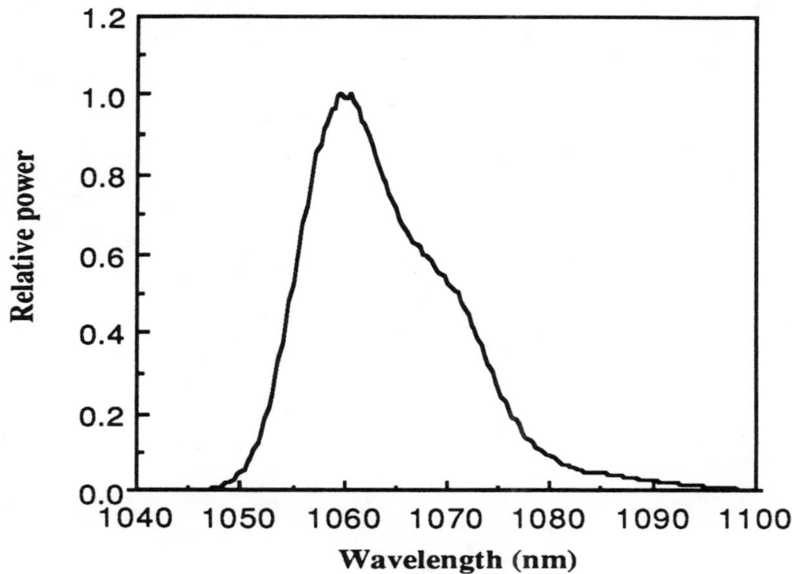

Fig. 4 Spectrum of 1060 nm line with 4.8 mW total power for pump wavelength of 825 nm

References

1. Bergh, R.A., Lefevre, H. C., and Shaw, H. J., "An Overview of Fiber-Optic Gyroscopes", IEEE Journal of Lightwave Technology, LT-2, 2, 91-107

2. Po, H., Hakimi, F., Mansfield, R.J., McCollum, B. C., Tumminelli, R. P., and Snitzer, E., "Neodymium Fiber Laser at 0.905, 1.06 and 1.3 microns, " Optical Society of America Annual Meeting, Seattle, Washington (October 1986).

3. Digonnet, M. J. F., "Theory of Superfluorescent Lasers," IEEE Journal of Lightwave Technology ,LT-4, 11, 1631-1639 (1986).

FIRDAY, JANUARY 29, 1988

MEETING ROOM 2-4-6

1:30 PM–3:00 PM

FEE1–5

POLARIZATION MAINTAINING COMPONENTS

Eric Udd, McDonnell Douglas Astronautics Company, *Presider*

Fiber Devices for Fiber Sensors

Juichi NODA and Itaru YOKOHAMA

NTT Opto-electronics Laboratories
Tokai-mura, Ibaraki-ken, 319-11 JAPAN
Tel 81-292-87-7523, Fax 81-292-87-7877

1. Introduction

Fiber devices are promising for many fiber sensor application systems. Before developing polarization-maintaining fibers, the functions for fiber devices are limited into a few kinds. Since polarization-maintaining fibers have been developed [1], many kinds of fiber devices have been proposed as new applications such as coherent optical transmission systems, fiber optic gyroscopes etc.. Fiber devices and their application reported upto now are summarized in Table 1. This paper reviews recent fiber devices mainly based on single-mode and polarization-maintaining fibers.

2. Fiber Devices

(1) COUPLERS: Fiber couplers (FC) are the most important devices. They are classified into two types; polishing types and fused-elongation types. The polishing types are advantageous for being able to be fabricated without taking into account the fiber structures. However, the coupler fabrication is difficult and the performance is unreliable for temperature fluctuation.

The fused elongation method is more practical than the polishing method due to easy fabrication, reliable characteristics and coupler varieties. Excess losses less than 0.1dB and high fabrication reliability are obtained by introducing automation processes [37,38]. When the coupling length is long, coupling power changes oscillatory to the wavelength. This function has the potential for multi/demultiplexers [7,8]. The other function is flat coupling chracteristics for broadband performance in wavelength by fusing the fibers with different parameters [9,10].

In case of polarization-maintaining (PM) FC fabrication, principal axis alignment for PM fibers is important. To suppress mis-alignment pronounced during the fusion elongation process, two micro burners set symmetrically are used [38]. Low crosstalk less than -30dB using PANDA fibers is achieved with these burners. Excess losses of the conventional PANDA FCs, however, are high due to the presence of index-depressed stress applying parts [11]. Two solutions have been proposed to reduce the excess losses. One solution is to use the PANDA fibers with index-matched stress applying parts [12]. We have recently found that the large diameter in the fused region is effective as an other solution while the coupling length becomes long [39]. Results of two solutions are summarized in Table 2. Low excess losses exhibit in all available wavelength region as shown in Fig.2. When the coupling length becomes longer by more forth times than usual lengh (-5mm), the PM FCs have polarization-splitting function as shown in Fig.3. The polarization splitter operates at 1.282um. 1.346um and 1.402um in wavelength.

(2) POLARIZATION DEVICES: There are two kinds of polarizers. One is made of metal loading or birefringent crystal loading on the core removed by polishing. In these types the transmitted mode also suffers high loss, although the extinction ratio more than 30dB is obtained [15]. The other type is a coiled PM fiber to produce different bending losses between orthogonal modes [18]. The spectral width of 340nm around 1.3um, the extinction ratio of 42 dB and the excess loss of 0.3dB are obtained with PANDA fibers, whose characteristics are almost invariable from -60°C to 100°C. Further improvement for being insensitive to side pressure and being small size in coil is necessary.

Fiber depolarizers using PM fibers have been well studied experimentally and theoretically by taking into account the mode coupling in PM fibers [19,40]. The azimuth input angle in applying the depolarizer to OTDR was found to be 27° and 63° [41] which are different from 45° used in conventional depolarizers. In case of depolarizing coherent light such as DFB lasers, the structure with a fiber delay line is proposed [21].

Polarization state controllers are constructed with single-mode fibers [22]. Their operations are obtained by using birefringence induced by bending, side-pressing, twisting or Farady effect. Most of their performances are limited in the range of phase control. To perform the endless operation two kinds of a rotatable fiber crank [22] and a improved rotatable coil are proposed [42].

(3) FILTERS: Six kinds of filters are proposed as seen in Table 1. Among them the ring resonators are most promising for appli cation to fiber gyroscopes or fiber lasers based on the filter function. Highest finnes of about F=500 is obtained using the single-mode fibers [28]. Key component for developing the ring resonatores is the high quality PM-FCs. Fabry-Perot resonators are also potential for the fiber lasers as well as the filters.

(4) OTHER DEVICES: Nonreciprocal devices such as isolators and circulators are useful for application to fiber sensors as well as optical transmission systems. However they are still unavailable for practical use. Fiber lasers or nonlinear fiber devices are potential for fiber sensors by using doped fibers [35,36] and/or FCs.

References

[1] J.Noda, K.Okamoto and Y.Sasaki: J. Lightwave Technol., LT-4, p146(1986).
[2] V.J.Tekippe and W.R.Willson: Laser Focus, May, 132(1985).
[3] C.M.Ragdale, M.M.Slonecker and J.C.Williams: SPIE,479, p2(1984).
[4] B.S.Kawasaki, M.Kawachi, et al: J.Lightwave Technol., LT-1, p176(1983).
[5] K.Imoto, M.Maeda, et al: J. Lightwave Technol., LT-5, p694(1987).
[6] D.B.Mortimore: Electron. Lett., 22, p1205(1986).
[7] D.B.Mortimore: Electron. Lett., 21. p742(1985).
[8] K.L.Sweeney, M.Corke, et al: SPIE, vol.630, 141(1986).
[9] C.M.Lawson, P.M.Kopera, et al: Electron. Lett., 20, p960(1984)
[10] G.Georgiou and A.C.Boucuvalas: Electron. Lett., 22, p62(1986).
[11] M.Kawachi, B.S.Kawasaki, et al: Electron. Lett., 18, p962(1982).
[12] I.Yokohama, K.Okamoto, et al: Electron. Lett., 22, p929(1986).
[13] I.Yokohama, K.Okamoto and J.Noda: Electron. Lett., 21, p415(1985).
[14] W.Eickhoff: Electron. Lett., 16, p762(1980).

[15] R.A.Bergh, H.C.Lefevre and H.J.Shaw: Opt. Lett., 5, p479(1980).
[16] V.P.Varnham, D.N.Payne, et al: Electron. Lett., 19, p246(1983).
[17] J.R.Simpson, R.H.Stolen, et al: J. Lightwave Technol., LT-1, p370(1983).
[18] K.Okamoto, T.Hosaka and J.Noda: J. Lightwave Technol., LT-3, p758(1985).
[19] K.Bohm, K.Peterman and W. Widel: J. Lightwave Technol., LT-1, p71(1983).
[20] W.K.Burns: J.Lightwave Technol., LT-1, p475(1983).
[21] K.Takada, K.Okamoto and J.Noda: J. Lightwave Technol., LT-4, p213(1986).
[22] T.Okoshi:J. Lightwave Technol., LT-3, p1232(1985).
[23] M.S.Yakaki, D.N.Payne and M.P.Varnham: Electron. Lett., 21, p248(1985).
[24] K.Okamoto and J.Noda: Electron. Lett., 22, p211(1986).
[25] Y.Yen and R.Ulrich: Opt. Lett., 6, p278(1981).
[26] K.Okamoto, J.Noda and H.Miyazawa: Electron. Lett., 21, p90(1985).
[27] R.H.Stolen, A.Ashkin, et al: Opt. Lett. 9, p300(1984).
[28] M.H.Yu and D.B.Hall: SPIE, 478, p104(1984).
[29] J.Stone: Electron. Lett., 21, p504(1985).
[30] I.Bennion, D.C.J.Reid: Electron. Lett., 22, p341(1986).
[31] K.Okamoto, T.Morioka, et al: Electron. Lett., 22, p181(1986).
[32] K.Okamoto, H.Miyazawa and J.Noda: Electron. Lett., 21, p36(1985).
[33] I.Yokohama, K.Okamoto and J.Noda: Electron. Lett., 21, p746(1985).
[34] I.Yokohama, K.Okamoto and J.Noda: Electron. Lett., 22, p370(1986).
[35] R.J.Mears, L.Reekie, et al: Electron. Lett., 21, p739(1985).
[36] M.Shimizu, M.Horiguchi and H.Suda: Electron. Lett., 23, p768(1987).
[37] M.Corke, M.Kale, et al: SPIE, 20, p963(1984).
[38] I.Yokohama, K.Okamoto and J.Noda: J. Lightwave Technol., LT-5, p910(1987).
[39] I.Yokohama, K.Chida and J.Noda: Appl. Opt. (to be submitted).
[40] K.Takada, K.Chida and J.Noda: Appl. Opt., 26, p2979(1987).
[41] T.Horiguchi, K.Suzuki, et all: J. Opt. Soc. Am. A, 2, p1698(1985).
[42] T.Matsumoto and H. Kano: Electron. Lett., 22, p78(1986).

TABLE 1 Classification of fiber devices

Category	Devices	Fibers			References
		MM	SM	PM	
Couplers	3dB or Accessing FC	○	○	○	[2]-[4]
	Star FC	○	○	—	[5],[6]
	Broadband FC	○	○	—	[7],[8]
	Multi/Demultiplexing FC	X	○	○	[9],[10]
	Polarization-Maintaining FC	X	X	○	[11],[12]
	Polarization-Splitting FC	X	X	○	[13]
Polarization Devices	Polarizer	X	○	○	[14]-[18]
	Depolarizer	X	—	○	[19]-[21]
	Polarization State Controller	X	○	—	[22]
Filters	Lyot Type Filter	X	○	○	[23],[24]
	Solc Type Filter	X	○	○	[25]-[27]
	Ring Resonator	X	○	○	[28]
	Fabry-Perot Filter	X	○	○	[29]
	Gratting Filter	X	○	○	[30]
	Birefringent Filter	X	○	○	[31]
Other Devices	Isolator	X	○	○	[32]
	Circulator	X	X	○	[33],[34]
	Laser	—	○	—	[35],[36]

FC : Fiber Coupler
MM : Multi Mode
SM : Single Mode
PM : Polarization-Maintaining
X : Impossible
— : Not Reported

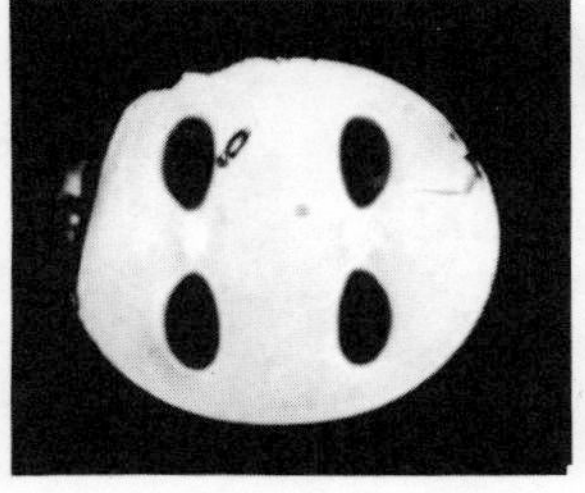

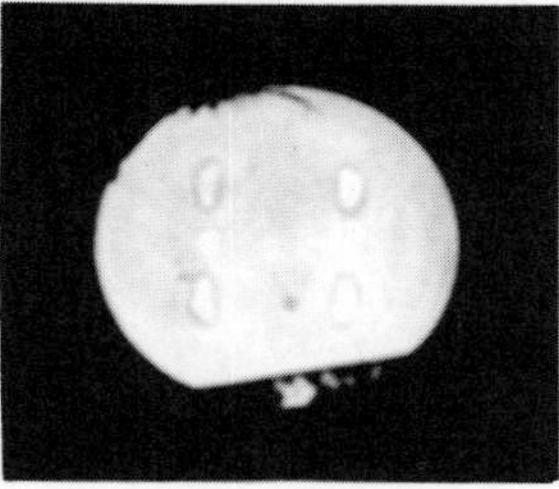

(a) P1 coupler (b) P2 coupler

Fig. 1 Cross sectional views of fused region. (a) P1 series are made of PANDA fibers with index-depressed stress applying parts. (b) P2 series are made of those with index-matched stress applying parts.

TABLE 2 Characteristics of P1 and P2 couplers

Coupler	Minimum Diameter (μm)	Excess Loss (dB)	Coupling Ratio (%)	Crosstalk (dB)
P1-1	60	0.05	38	−31.9
P1-2	51	0.9	1	−27.3
P1-3	41	1.7	99	−26.2
P1-4	32	1.2	33	−27.5
P2-1	52	0.03	43	−32.1
P2-2	34	0.15	22	−25.4
P2-3	25	0.18	1	−28.9

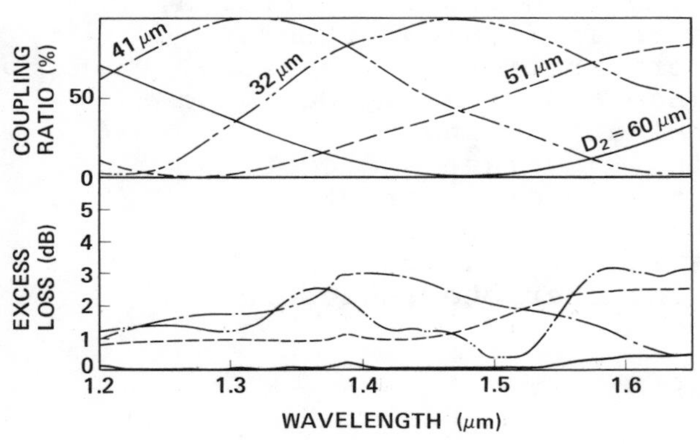

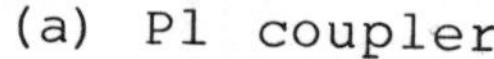

(a) P1 coupler

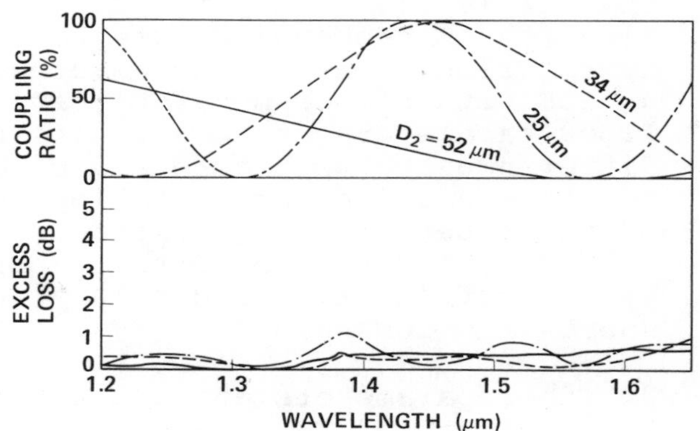

(b) P2 coupler

Fig. 2 Spectral characteristics of excess losses and coupling ratios.

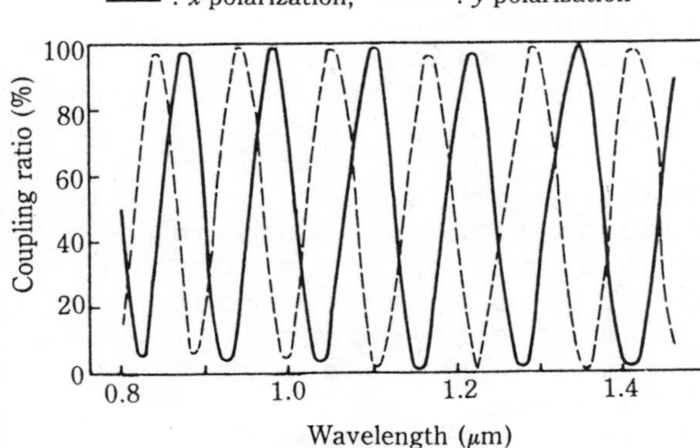

Fig. 3 Spectral characteristics of coupling ratios for PANDA fiber coupler with long coupling length. Coupling length is about 20 mm.

A HIGH-EXTINCTION-RATIO AND LOW-LOSS SINGLE-MODE SINGLE-POLARIZATION OPTICAL FIBER

K. HIMENO, Y. KIKUCHI, N. KAWAKAMI, O. FUKUDA, and K. INADA

FUJIKURA LTD.
1440 MUTSUZAKI, SAKURA-SHI, CHIBA-KEN, 285 JAPAN
TEL. : 0434-84-2111 FAX. : 0434-86-8561

1. Introduction

Single-mode single-polarization optical fibers (SPFs) /1/ which can guide only one polarization mode have many applications in polarizers/2/, fiber sensors, and coherent-transmission lines.

In those applications, high-extinction ratio and low-loss SPFs are required. In short-length operation such as polarizer, SPFs are required to keep a high extinction ratio and a low-loss, when the fibers are coiled in small size. The bending performance of the guided polarization mode of the SPF is also a key to achieving the long-length operation, since the microbendings are occur even in straight fibers by the fiber coating and external perturbations. Therefore, an SPF with higher modal birefringence and properly designed fiber parameters such as relative refractive index differences between core and cladding, V values and coating structures is requested to improve the bending performance of the fiber as a high extinction ratio is kept./3/,/4/

This paper reports the fabrication of low-loss SPF with birefringence of 8×10^{-4} and the characteristics in short- and long-length.

2. Fiber parameters and characteristics

Fig. 1 shows the structure of the PANDA /5/ fiber. The fiber has been fabricated by a pit-in-jacketing method. To achieve an SPF with high birefringence, the normalized distance between stress applying parts (SAPs), c/b was designed less than 0.1, and B_2O_3 mol concentration of SAPs was more than 18mol%. Core-cladding relative refractive index difference Δn and core diameter 2a was designed for the operation wavelength in 0.85um wavelength band. Modified silicone was used for the primary coating, because the UV cured coating easily induces microbendings and increase polarization crosscoupling. The fiber parameters are shown in Table 1.

Fig. 2 shows loss spectrum of the fiber. The differential loss between the two orthogonal modes occurs in 0.85μm band. This fact shows single-polarization operation is achieved in 0.85μm band. The loss of X-polarization mode is 2.7dB/Km, which is comparable to that of conventional single-mode fibers, although OH-absorption loss appears at 0.88μm because of lower c/a and lower V value.

Bending losses of two orthogonal modes were measured at

0.85μm wavelength using an LD light source and a Gran-Thompson prism with extinction ratio of 60 dB. The results are shown in Fig. 3. From this measurement, each bending loss of the two orthogonal modes of the fiber is described as

$$\log \alpha x = -0.118D + 3.59$$

$$\log \alpha y = -0.0455D + 2.92$$

where αx , αy are X-polarization and Y-polarization mode losses in decibels, respectively, and D is a bending diameter in mm.

Those results prove that greater birefringence increases the differential loss. As Δn and 2a are designed properly, the bending loss of X-polarization modes seems to be negligibly small in the bending diameter over 30mm. Therefore, it is possible to widen the choice of coiling diameter. For example, the fiber bending diameter is from 60mm to 80mm to obtain a 500m length of SPF with an extinction ratio more than 40dB without increasing X-polarization mode loss. Such an SPF with a small X-polarization loss keeps its performance against external perturbations.

A fiber polarizer and SPF coils were fabricated using the fiber. Those characteristics are shown in Table 2. The crosstalk η_x or η_y was measured while X- or Y-polarized light is launched, respectively. The extinction ratio was the power ratio of X- and Y-polarized lights at the fiber output while circularly polarized light is launched at the fiber input. In both short- and long-length, good performance is obtained;the differential modal loss is more than 53dB and the extinction ratio is more than 47dB for each operation. Low polarization crosscoupling can be achieved because of few internal perturbations like core and/or SAPs deformation. But such low polarization crosscoupling, as shown with a 1500m fiber in Table 2, cannot be realized by using conventional polarization-maintaining fiber. It is clear that the differential loss minimizes the polarization crosscoupling.

3. Conclusion

A Highly birefringent PANDA fiber for single-mode single polarization operation has been fabricated. A fiber with modal birefringence of 8×10^{-4} exhibits low-loss of 2.7dB/km at 0.85μm wavelength and an extinction ratio of 47dB in short- and/or long-length. Such a good performance can be achieved by increasing the modal birefringence and reducing the internal perturbations.

4. References

/1/ J. R. Simpson, et al., J. Lightwave Technol., 1983, JLT-1, pp. 370-374
/2/ M. P. Varnham, et al., Opt. Lett., 1984, 9, PP. 306-308
/3/ M. J. Marrone, Electron. Lett., 1985, 21, pp. 244-245
/4/ K. Okamoto, et al., J. Lightwave Technol., 1985, LT-3, pp. 758-762
/5/ Y. Sasaki, et al., Electron. Lett., 1983, 19, pp. 792-794

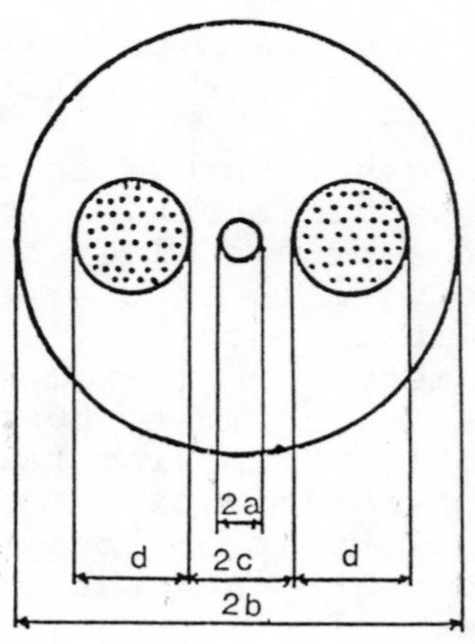

Fig. 1 Structure of a PANDA fiber

Table 1 Fiber parameters

Δ n (%)	0.34
Core diameter 2a (μm)	3.6
Fiber diameter 2b (μm)	125
Distance between SAPs 2c (μm)	6.2
SAP diameter d (μm)	35
Cutoff wavelength (μm)	0.55
Modal birefringence	8.6×10^{-4}
Crosstalk (dB/5m)	-47
Primary coating	Modified Silicone 200um
Buffer coating	UV curable Epoxy Acrylate 250um

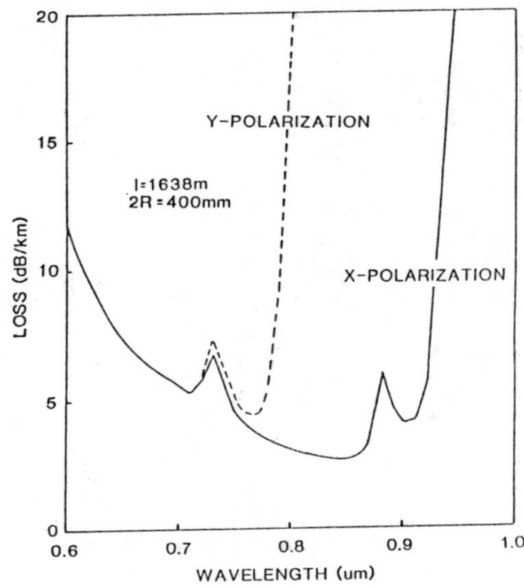

Fig. 2 Loss spectrum of the SPF

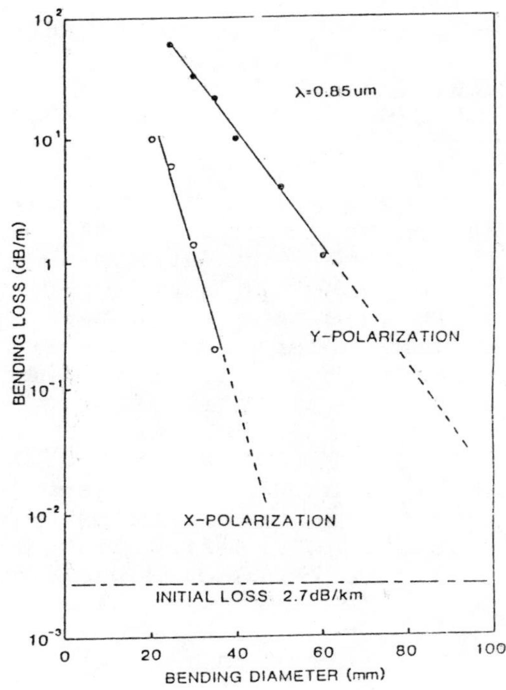

Fig. 3
Bending losses of two orthogonal polarized modes of the SPF vs. bending diameters

Table 2 Characteristics of SPFs

	No.1	No.2	No.3
Bending diameter (mm)	35	80	400
Length (m)	3.9	500	1500
Crosstalk η_x (dB)	-49	-48	-48.5
Crosstalk η_y (dB)	4.8	5.1	4.1
Extinction ratio (dB)	48.1	47.0	47.1
Loss α_x (dB)	0.3	2.6	4.1
Loss α_y (dB)	54	55.6	57.4
Differntial loss $\alpha_y - \alpha_x$ (dB)	53.5	53.0	53.3

HIGH PERFORMANCE POLARIZERS AND SENSING COILS WITH ELLIPTICAL JACKET TYPE SINGLE POLARIZATION FIBERS

Y.TAKUMA, H.KAJIOKA AND K.YAMADA

HITACHI CABLE, LTD.

319-14 5-1-1 HITAKA-CHO, HITACHI-SHI, JAPAN
TEL 0294-42-3151 (EXT) 3692

1. INTRODUCTION

Single polarization fibers which guide only one polarization mode would be useful in fiber optic sensors/1/. It was shown by Varnham et al/2/ that the polarizing effect can be enhanced when the birefringent fiber is bent.

In this paper we present experimental investigations on in-line polarizers and sensing coils for fiber optic gyroscope that we have developed using elliptical jacket type single polarization fibers.

2. HIGH PERFORMANCE POLARIZERS

Fig.1 shows a cross-sectional view of elliptical jacket type single polarization fiber and its refractive index profile. This fiber is composed of four regions: a concentric circular GeO_2 doped core and silica cladding which compose a low loss waveguide, and a B_2O_3/P_2O_5 doped elliptical jacket and silica outer support to produce a large anisotropic stress in the core. The core and the cladding is fabricated by VAD method and the preform by Rod-in-Tube method. To realize a high performance polarizer, the fiber is designed so that the two orthogonal modes of polarizations have different attenuation due to winding. Therefore the elliptical jacket is designed to be depressed and the normalized frequency is made as low as sufficient.

Fig.2 shows the bending loss characteristics of the orthogonal modes of polarizations. By optimizing fiber parameters we have developed high performance polarizers as summarized in Table 1. These data show that this in-line device can be substituted for bulk polarizers in fiber optic sensor system that use high birefringent fiber.

Fig.3 shows a schematic diagram of an all-fiber gyroscope(based on the phase modulation method). Fig.4 shows the effect of the polarizer. The parameters of the trial fiber optic gyroscope are as follows. The diameter of the sensing coil is 80 mm and the fiber length is 400 m. The fiber is conventional elliptical jacket type polarization maintaining fiber/3-4/. The optical source is RF-modulated laser diode with wavelength 0.85 µm. It is obvious that this device can effectively compensate for divergences from a true rotation rate that is caused by phase bias due to the finite extinction ratios/5/ of components of the fiber optic gyroscope.

3. HIGH PERFORMANCE SENSING COIL

By optimizing fiber parameters we have developed a sensing coil with a diameter of 80 mm which has an extinction ratio of -45 dB/0.5 km with no excess loss. The principle of this device is same as that of polarizers with elliptical jacket type single polarization fibers.

Fig.5 shows the temperature characteristics of this sensing coil. An extinction ratio of -45 ±0.5 dB was obtained for a temperature range of -20 to 40 °C. A major advantage of this new sensing coil for fiber optic gyroscope applications is that it is effective at reducing the polarization fluctuation induced drift of conventional sensing coils.

4. CONCLUSION

In-line polarizers using elliptical jacket type single polarization fibers with extinction ratio of less than -40 dB and insertion loss of less than 0.5 dB have been developed. A sensing coil with extinction ratio of -45 dB/0.5 km and no excess loss has also been developed. These devices will prove to be useful not only in fiber optic gyroscope but in other sensor and coherent communications systems as well.

REFERENCES

/1/ K.Okamoto et al, Electron. Lett., 20, p429, 1984.
/2/ M.P.Varnham et al, Electron. Lett., 19, p679, 1983.
/3/ H.Matsumura et al, 6th ECOC(York), p49,Sept.,1980.
/4/ H.Kajioka et al, 8th ECOC(Cannes), p143, Sept., 1982.
/5/ E.C.Kinter, Opt. Lett., 6, p154, 1981.

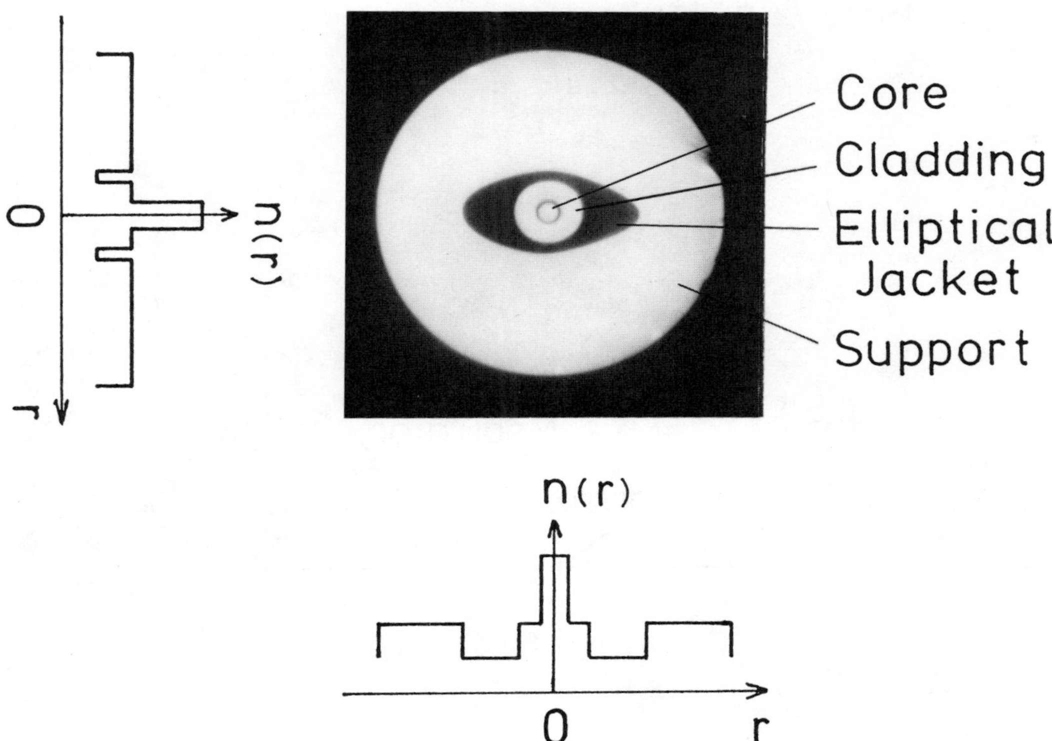

Fig.1 Cross-sectional view of elliptical jacket type single polarization fiber and its refractive index profile.

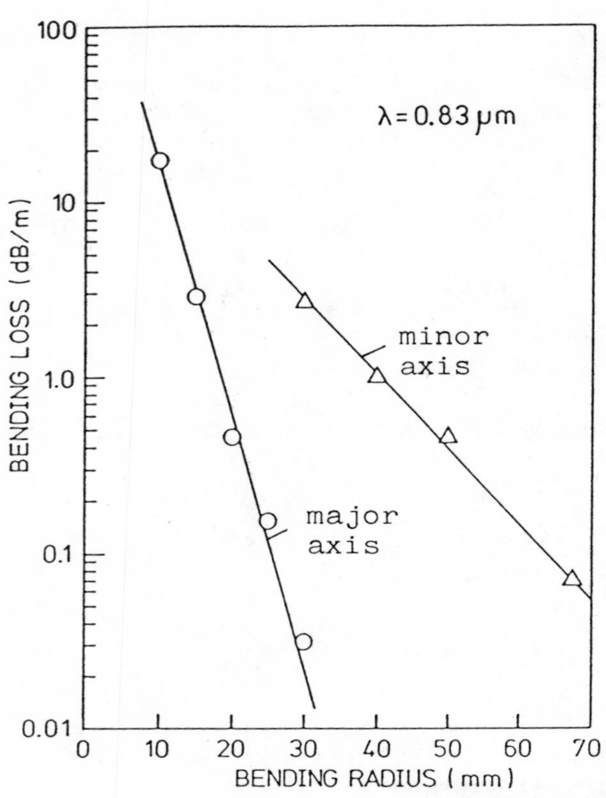

Fig.2 Bending loss characteristics

Table 1 Characteristics of in-line polarizers

ITEMS / WAVE-LENGTH	BENDING RADIUS (mm)	NUMBER OF TURNS	EXTINCTION RATIO (dB)	INSERTION LOSS (dB)
0.85 μm	40	15	-42	0.3
	50	18	-44	0.2
	60	20	-45	0.1
1.30 μm	40	15	-42	1.0
	50	35	-42	0.9
1.55 μm	40	40	-40	0.5

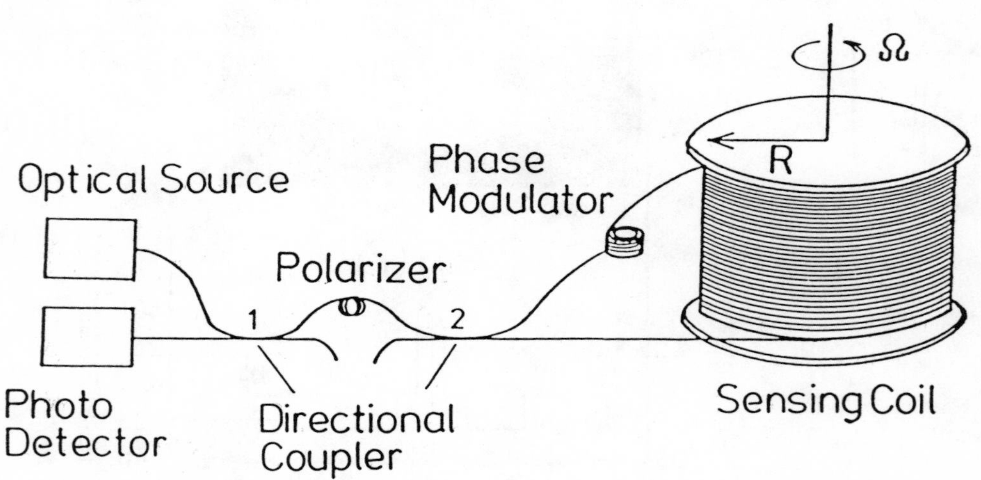

Fig.3 Basic set-up of all-fiber gyroscope (phase-modulation method)

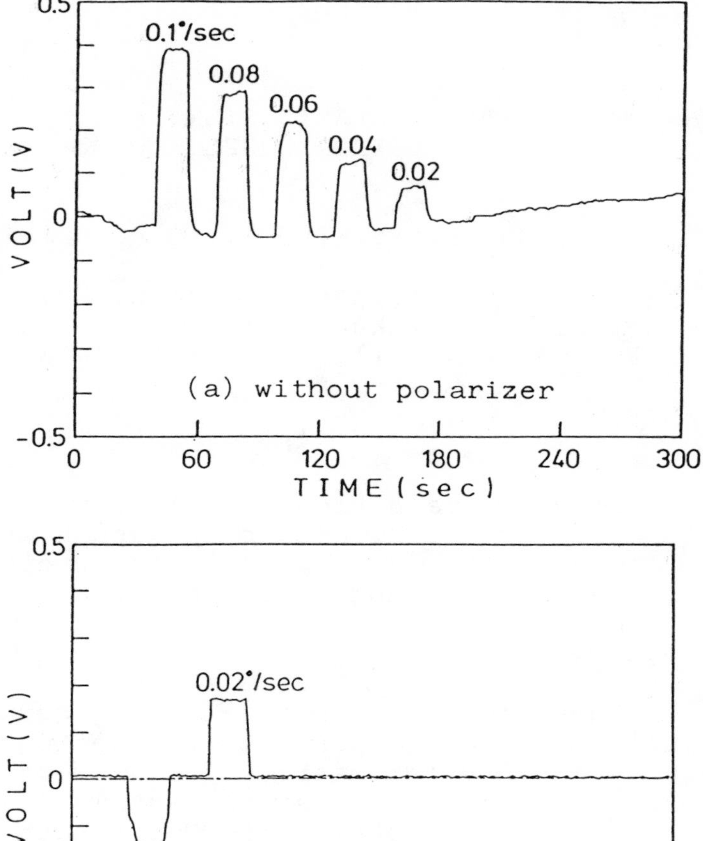

Fig.4 Effect of in-line polarizer in fiber optic gyroscope. Bending radius and number of turns of this polarizer are 41 mm and 15, respectively.

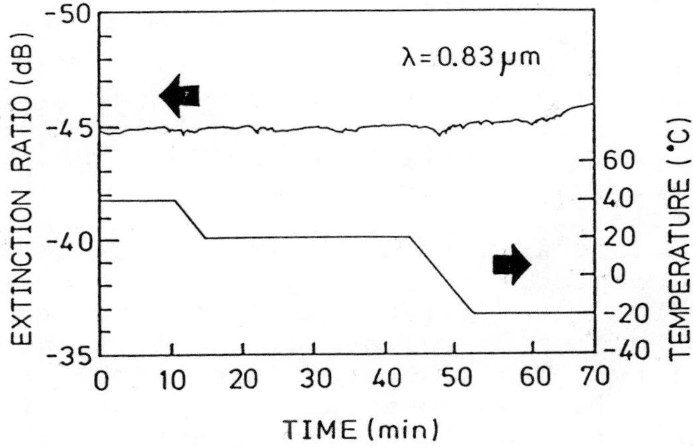

Fig.5 Temperature characteristics of sensing coil

ULTRA-LOW-CROSSTALK POLARIZATION MAINTAINING OPTICAL FIBER COUPLER

T.ARIKAWA, F.SUZUKI, Y.KIKUCHI, O.FUKUDA and K.INADA

Research and Development Division, Fujikura Ltd.,
1440, Mutsuzaki, Sakura-shi, Chiba, 285, JAPAN
Tel : 0434-84-2111 Fax : 0434-86-8561

Abstract

The experimental evaluation of polarization-maintaining optical fiber couplers fabricated by polishing technique is described. The couplers show polarization crosstalk of less than -30dB and small wavelength dependence of splitting ratio.

1.Introduction

Polarization-maintaining optical fiber couplers (PMCs) [1] are important as optical compornents for the optical fiber sensors such as an interferometric optical fiber sensor and for coherent optical fiber communication systems [2].

For the fabrication methods of PMC, fusion-elongation method [1][3] and mechanical polishing method [4][5] are well known. In conventional fabrication of fused coupler, the principal axis of a polarization-maintaining fiber rotates in fusion-elongation process. Hence, the polarization crosstalk of a coupler is degraded by this misalignment. In this paper, PMCs made by polishing method are investigated. This method is effective to avoid the misalignment degrading the polarization crosstalk. Two experimental evaluations are mainly described. At the beginning, the relation between polarization crosstalk and the misalignment of polarization-principal axes of fibers is described in detail. Then, the wavelength dependence of splitting ratio is investigated for each wavelength of 0.85, 1.3 and 1.55 um.

2.Experimental

PMCs for use at each wavelength of 0.85, 1.3 and 1.55 um were fabricated using stress-induced type polarization-maintaining fibers so called PANDA (Polarization And Absorption reducing) fibers[3]. Fiber characteristics and parameters are listed in Table 1.

Fig.1 illustrates two arrangements in cross section of PANDA fiber coupler. The arrangement of case (a) in this figure was adopted for coupler fabrication for the reason that low polarization crosstalk values were easily achieved in good reproducibility. The reason of this is that the polarization principal axis is aligned accurately in case (a) when a PANDA fiber is fixed into a groove of a quartz block [5].

Fig.2 shows the cross-sectional transverse view of a PMC for use at wavelength of 1.3 um. Polarization crosstalk of many couplers were measured on coupled light in couplers. The relation between the polarization crosstalk (CT) of coupled power and the misalignment angle ($\Delta\theta$) of the polarization-principal (slow) axis can be approximated by using the equation (1). Polarization

$$CT=\tan^2(\Delta\theta) \qquad -(1)$$

crosstalk were evaluated experimentally and theoretically.

3.Result

Fig.3 shows the experimental result of the polarization crosstalk as a function of polarization axes misalignment at the wavelength of 1.3 um. In Fig.3, the circles show the polarization crosstalk of many couplers and the curved line shows the equation (1). From this figure, it can be seen that the experimental results correspond to the equation (1) so closely. It is assumed that a little irregularity of experimental results is caused by the stress of adhensive which is used to fix a fiber into a groove of a quartz block. It is cleared that polarization crosstalk is correlated with the principal axis misalignment. Consequently, it can calculate that the $\Delta\theta$ should be keep less than 1.8 degrees in order to obtain the polarization crosstalk less than -30dB.

Table 2 shows the characteristics of the PMCs which were fabricated controlling the $\Delta\theta$ so little. The PMCs for use at each wavelength of 0.85, 1.3 and 1.55 um show good characteristics that the polarization crosstalk is better than -30dB and the excess loss is less than 0.5dB.

Fig.4 shows the wavelength dependence of splitting ratio for three PMCs listed in Table 2. As shown in this figure, the wavelength dependence of splitting ratio were about 2%/10nm for CPL-SM85P and about 1%/10nm for CPL-SM13P and CPL-SM15P, and were relatively smooth. The wavelength dependence of splitting ratio were tend to be larger as the wavelength became shorter.

4.Conclusion

The correlation between the principal axes misalignment and the polarization crosstalk of PMCs fabricated by machanical polishing method has been cleared experimentally. Couplers produced by accurate polarization-principal alignment exhibit ultra-low-crosstalk characteristics of less than -30dB. Furthermore, these PMCs have low excess loss less than 0.5dB and small wavelength dependence of splitting ratio. They are applicable for the interferometric optical fiber sensors and coherent optical fiber communication systems.

References

[1]M.KAWACHI, et al. : Electron. Lett., Vol.18, pp 962-964, 1982
[2]J.NODA, et al. : J. Lightwave Technol., Vol.LT-4 pp 1071-1089, 1986
[3]I.YOKOHAMA, et al. : Electron. Lett., Vol.22, pp 929-930, 1986
[4]W.PLEIBEL, et al. : Electron. Lett., Vol.19, pp 825-826, 1983
[5]F.SUZUKI, et al. : Microoptics Conference '87, TOKYO, 1987

Table 1 Characteristics and parameters of PANDA fibers

Item \ Fiber	SM85-P	SM13-P	SM15-P
Wavelength (μm)	0.85	1.30	1.55
Crosstalk (dB)	-38	-39	-37
Loss (dB/km)	2.4	0.5	0.4
Core dia. (μm)	5	8	9
Fiber dia. (μm)	125	125	125
Coating dia. (mm)		0.4	
Material		UV Cured resin	

Table 2 Characteristics of PANDA fiber couplers

Item \ Coupler		CPL-SM85P	CPL-SM13P	CPL-SM15P
Wavelength (μm)		0.85	1.30	1.55
Crosstalk (dB)	1→3	-34	-35	-33
	1→4	-33	-34	-33
Excess loss(dB)	1→3	0.5	0.3	0.5
	1→4	0.4	0.4	0.5
Splitting ratio (%)		51	52	49
Isolation (dB)		50	51	50

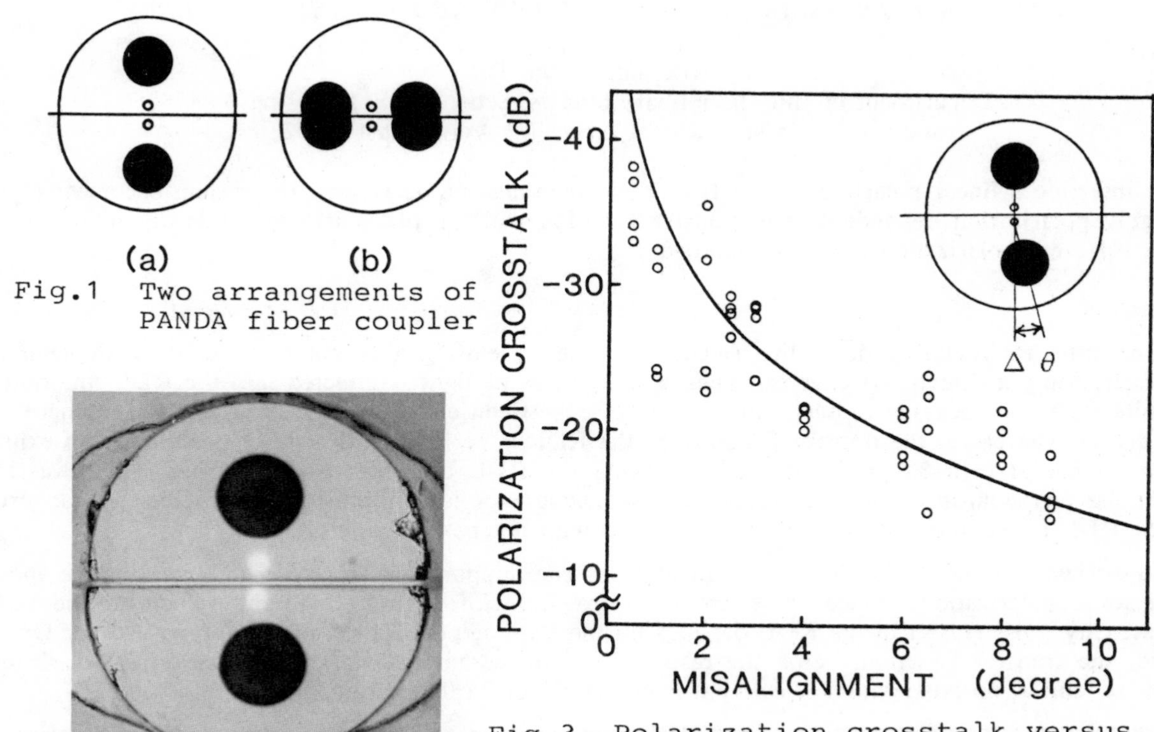

Fig.1 Two arrangements of PANDA fiber coupler

Fig.2 Cross-sectional view of PANDA fiber coupler

Fig.3 Polarization crosstalk versus misalignment between principal axes

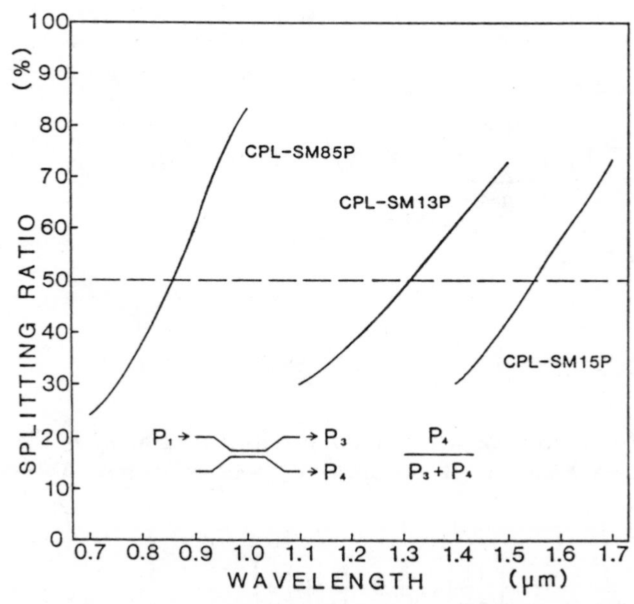

Fig.4 Wavelength dependence of the splitting ratio of PANDA fiber couplers

POLARIZATION CONTROLLED FIBER OPTIC RECIRCULATING DELAY LINE FILTER AND ITS ASSOCIATED PHASE INDUCED INTENSITY NOISE

Ady Arie and Moshe Tur
Department of Interdisciplinary Studies, School of Engineering
Tel Aviv University, Ramat Aviv, Tel Aviv, Israel 69978

Abstract

By inserting a linear polarizer after a fiber optic recirculating delay line, the system becomes highly polarization dependent. The transfer function and the phase induced intensity noise for different polarization inputs are studied.

1. Introduction

The fiber optic recirculating delay line (RDL) is made by closing a continuous strand of single-mode fiber on itself, using a directional coupler. When a short pulse of light is injected into the RDL, the output is an infinite series of decaying pulses, with a delay of τ between every two consecutive pulses (where τ is the RDL delay). Therefore, the transfer function of the RDL is periodic with a $1/\tau$ period, and can exhibit very deep notches at $f=(n+0.5)/\tau$ for n=0,1,2.., making the RDL useful as a notch filter [1]. It was also shown that the phase noise of the short coherence semiconductor laser illuminating the fiber, is converted through the RDL to spectrally structured intensity noise with notches at $f=n/\tau$ [2].

A non-polarization holding single-mode optical fiber can support two approximately degenerate modes with orthogonal polarizations. Since these modes propagate with slightly different velocities, the phase induced intensity noise (PIIN) in the RDL depends on the state of polarization of the input field [2]. On the other hand, the transfer function, being the result of an incoherent addition of intensities, is polarization insensitive, as long as no polarization selective elements are used in the circuit.

Here we suggest a modified version of the RDL (Fig. 1), with an output polarizer and two polarization controllers, to control the input polarization, as well as the polarization of the recirculating field. It will be shown that by proper adjustment of the polarization controllers, the filtering notches at $f=(n+0.5)/\tau$ are replaced by notches at $f=(n+0.5)/2\tau$. This adjustment also has a profound effect on the shape of the spectrum of the PIIN. After summarizing the essential theoretical considerations in Sec. 2, the experimental results are presented and discussed in Sec. 3.

2. Theoretical Analysis

The output field of a fiber optic recirculating delay line is [2]

$$E_4(t)=\sum_{n=0}^{n=\infty} F_n \exp(i(w_0(t-n\tau)+\phi(t-n\tau))), \tag{1}$$

where the amplitude coefficients F_n are

$$F_n=\begin{cases}\delta_0 C & n=0\\ \delta_0 DB^{-1}A(\delta_0 B\exp[-\alpha_0 L])^n & n\geq 1\end{cases} \tag{2}$$

L is the loop length, α_0 the fiber amplitude attenuation per unit length, δ_0 is the coupler power insertion loss, and A,B,C,D are the complex elements of the unitary coupling matrix. The coupler is thus described by

$$\begin{bmatrix}E_3\\E_4\end{bmatrix}=\delta_0\begin{bmatrix}A & B\\C & D\end{bmatrix}\begin{bmatrix}E_1\\E_2\end{bmatrix} \tag{3}$$

(E_1, E_2) and (E_3, E_4) are, respectively, the input and output fields (see Fig. 1). The intensity impulse response is $\Sigma_n|F_n|^2\delta(t-n\tau)$, and its Fourier transform is the transfer function, $\Sigma_n|F_n|^2\exp(-2\pi i f n\tau)$. (Since we analyze the filter response of the RDL, we are assuming here total incoherence of the optical field). It is easily seen that the square modulus of the transfer function gets its maxima at $f=n/\tau$, and its minima at $f=(n+0.5)/\tau$ with deep notches, as long as the coupling coefficient is properly adjusted. For example, in a lossless loop, by adjusting the intensity loop transmittance to be 1/3, we get zero value of the transfer function at $f=(n+0.5)/\tau$.

If the system output consists of the odd pulses only (The series of the 1,3,5,...,2n-1, pulses from the original series), the situation is similar to the previous case, but the delay between successive pulses becomes 2τ. The period of the transfer function is halved, thus the maxima are obtained at $f=n/2\tau$. Note that these frequencies include a subgroup, $f=(n+0.5)/\tau=(2n+1)/2\tau$, at which the minima were obtained in the previous case. Therefore, by selecting the odd pulses, the deep notches of the square modulus of the transfer function at frequencies $f=(n+0.5)/\tau$ are replaced by the maxima of the function, while new notches appear at frequencies $f=(n+0.5)/2\tau$.

We shall evaluate now the PIIN of the system. It was shown in [3] that if we take the three first pulses of a recirculating delay system, we get an autocovariance function with a positive peak at t=0 together with two additional peaks at $t=\pm\tau$. The sign of these peaks depends on the term $\cos(\arg(F_0F_1^*)-\arg(F_1F_2^*))$, where F_0, F_1 and F_2 are the amplitude coefficients of the first three pulses, as defined in Eq. 2, and F^*_j denotes the complex conjugate of F_j. For the original RDL series this term is $\cos(\pi)=-1$, therefore the autocovariance function consists of a zero (positive) argument and two **negative** peaks at $t=\pm\tau$. Consequently, the PIIN, which is the Fourier transform of the autocovariance, gets its minima at $f=n/\tau$. If we select only the odd pulses, the sign of the peaks is determined by $\cos(\arg(F_0F_2^*)-\arg(F_2F_4^*))=\cos(\pi)$, thus we get again minima for the PIIN, but this time at $f=n/2\tau$. However, if we select the even pulses, i.e. the 2,4,6,.. pulses, the sign of the peaks is determined by $\cos(\arg(F_1F_3^*)-\arg(F_3F_5^*))=\cos(0)=+1$. Therefore the autocovariance function consists of a zero argument and two **positive** peaks. Consequently, the PIIN exhibits **maximum** noise, instead of **minimum** noise, at $f=n/2\tau$. Similar phenomenon, of PIIN spectrum "inversion" was also observed with a transmitting Fabry-Perot interferometer [4].

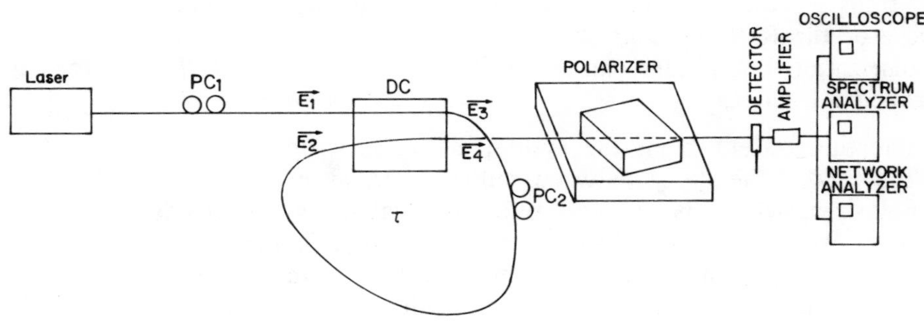

Fig. 1: A recirculating delay line with a polarizer and two polarization controllers, PC_1 and PC_2. DC the directional coupler and τ the loop delay. The impulse response is measured by the oscilloscope, the transfer function by the network analyzer, and the power spectrum of the phase induced intensity noise by the spectrum analyzer.

3. Experimental Results

The selection of odd and even terms can be accomplished with the fiber circuit of Fig. 1. The output of the 0.8μ semiconductor laser is fed into a single-mode fiber, with a polarization controller, PC_1, on it. The RDL itself is constructed of a directional coupler, a $\simeq$6 m loop of fiber and a second polarization controller, PC_2, which controls the polarization inside the RDL. An in line fiber optic polarizer[5] is placed after the RDL. The polarizer transmits linearly polarized light coming in the *a* direction, and blocks the light polarized in the orthogonal *b* direction. The output light intensity is detected by a square law photodetector. The system is analyzed in several interesting states of operation:

A. If all the optical fields after the RDL are in polarization *a*, they are not affected by the polarizer. This is equivalent to the case of an RDL alone, since the polarizer itself does not have any effect on the output fields. In order to achieve this state we first set the coupler to maximum coupling ratio between the input field, E_1, and the output field E_4, forcing all the light to pass through the coupler without entering into the RDL. Then, using PC_1 we set the optical field reaching the polarizer to polarization *a*. When the coupling ratio of the coupler is changed to full coupling between E_1 and E_3, all the light coming from the laser enters the RDL, circulates once through the loop, and comes out of the RDL into the polarizer. Using PC_2, we control the output field from the RDL to reach the polarizer in polarization *a*, too. Finally, we change the coupling ratio of the coupler to an intermediate state. If the coupler is polarization-independent, the field reaching the polarizer is always in the *a* direction, in which case the polarizer is transparent. As a result, one obtains , Fig. 2a, the same transfer function and PIIN as for the RDL without an output polarizer, Fig. 2e.

B. The impulse response of the RDL alone is an infinite series of pulses with a delay τ between successive pulses. If we tune the polarization controllers in the system, we can get a series in which the first pulse is in polarization *a*, the second pulse in polarization *b*, the third pulse in *a*, etc. The polarizer will block all the pulses in polarization *b*, leaving only the odd pulses from the original series. This case was analyzed in Sec. 2.

In order to obtain this situation we first check, as in case A, that the polarization of the first pulse reaches the polarizer in polarization *a*. Then, with full coupling between E_1 and E_3, PC_2 sets the RDL output field to the orthogonal *b* polarization (zero output). Actually, due to the loop and PC_2, the field is rotated by 90^0. When the coupling ratio is changed to an intermediate value, we get a series of pulses, alternating between the two orthogonal polarizations. However, only the *a* polarization can pass through the polarizer. The system output is similar to that of an RDL alone, but since the system delay is 2τ, the period of both the transfer function and the PIIN spectrum is halved (Fig. 2b). Since all even pulses are blocked by the polarizer, the transfer function can not approach zero at any frequency, thus the obtained notches are less deep than those of the RDL without polarizer. The results of cases A and B show that using the set-up of Fig. 1, the polarization controllers can control the shape of the RDL transfer function, making it a polarization controlled RF filter.

C. Another interesting output from the system is the series of the 2,4,6,..2n, pulses from the original output series of the RDL. The delay between pulses is 2τ, as in case B, but this time, the **first** pulse is blocked by the polarizer, while the **second** pulse is rotated by 90^0 inside the RDL loop, and can pass through the polarizer. We first bypass the loop and set with PC_1 the optical field at the polarizer input to polarization *b*. The RDL loop remains in the same condition as in case B, rotating the field polarization by 90^0 with every circulation. The results are obtained with the coupler in an intermediate state. The transfer function of this system (Fig. 2c) is similar to the one obtained in case B, since the impulse response is an infinite series with delay of 2τ between successive pulses. As explained in Sec. 2, the PIIN exhibits **maximum** noise at frequencies f=1/(loop delay), instead of **minimum** noise.

D. Suppose that the light that comes directly from the laser, bypassing the loop, is blocked by the polarizer, as in case C. However, the output from the RDL, depending on PC_2, is an arbitrary mixture of the two polarizations. The impulse response of the system can be an infinite series of pulses with a delay of τ between them, but the first pulse of the original RDL series is missing. The transfer function in this case is quite similar to that of RDL alone, with notches at $f=(n+0.5)/\tau$, but the power spectrum of the PIIN is "inverted", i.e. we obtain **maximum** noise at $f=n/\tau$.

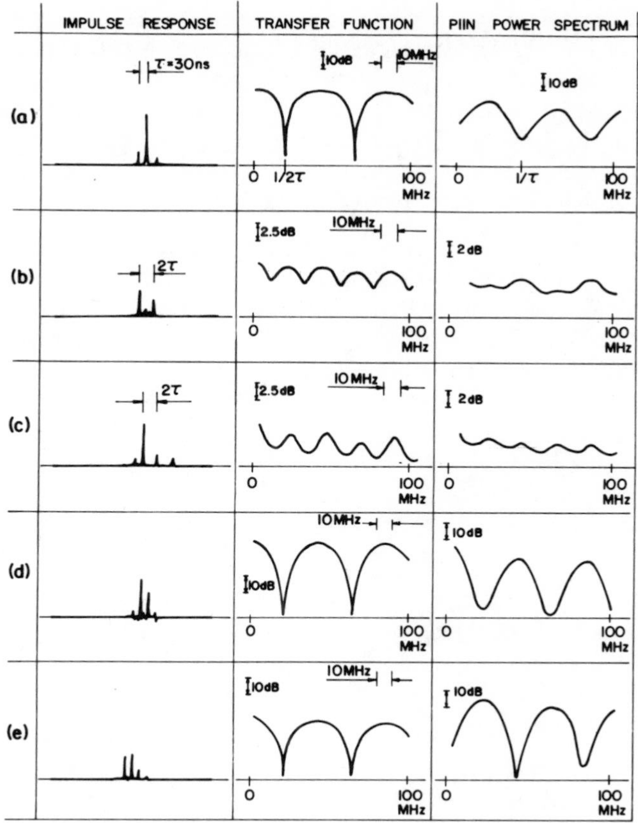

Fig. 2: Experimentally measured intensity impulse response (left), intensity transfer function (middle), and noise power spectrum (right) in different states of the system: a: case A. b: case B. c: case C. d: case D. e: RDL without a polarizer.

4. Conclusions

The dependence of the RDL and polarizer system on the input polarization and on the way by which it is changed through the RDL, may affect considerably the shape of the transfer function and the PIIN spectrum. The period of both functions may be halved, and the PIIN spectrum may exhibit **maximum** noise instead of **minimum** noise at frequencies f=n/(loop delay). An important result is that by controlling the polarization of the field, the depth of filtering notches at frequencies $f=(n+0.5)/\tau$ can be varied, without changing the coupling coefficient of the coupler, thus enabling the RF filter to be polarization controlled.

We would like to thank H. Mori and N. Konforty for their technical assistance and E.L. Goldstein for his helpful comments.

References

1. J.E. Bowers, S.A. Newton, W.V. Sorin and H.J. Shaw, "Filter Response of Single-Mode Fiber Recirculating Delay Lines", Electr. Lett., Vol. 18, p. 110-111, 1982.
2. M. Tur, B. Moslehi and J.W. Goodman, "Theory of Laser Phase Noise in Recirculating Fiber Optics Delay Lines", IEEE J. Lightwave Tech. vol LT-3, no.1, pp. 20-31, 1985.
3. M. Tur and B. Moslehi, "Laser Phase Noise Effects in Fiber Optic Signal Processing with Recirculating Loops", Opt Lett., vol. 8, pp 229-231, 1983.
4. E. Shafir and M. Tur, "Phase Induced Intensity Noise in an Incoherent Fabry-Perot and other Recirculating Devices", J. Opt Soc. Am., Vol. 4, No. 1, pp. 77-81, 1987.
5. R.A. Bergh, H.C. Lefevre and H.J. Shaw, "Single-mode Fiber Optic Polarizer", Opt. Lett. 5, pp. 479-481, 1980.

FRIDAY, JANUARY 29, 1988

MEETING ROOM 5-7-9

1:30 PM–2:45 PM

FFF1–5

TEMPERATURE SENSORS

Gordon Mitchell, MetriCor, Inc., ***Presider***

FFF1-1

PHASE-MEASUREMENT BASED RUBY FLUORESCENCE FIBRE OPTIC TEMPERATURE SENSOR

K.T.V. Grattan, R.K.Selli, A.W. Palmer
Measurement and Instrumentation Centre
School of Electrical Engineering and Applied Physics
City University, Northampton Square
London EC1V 0HB England.
Telephone: +44 1253 4399

INTRODUCTION

The authors have discussed in previous publications and a recent review [1] a number of the approaches which have been published in the literature on the design of fibre optic temperature sensors. There is considerable market potential for point sensors and the use of fluorescence-based techniques have been seen to be promising in both commericalized devices (e.g. Asea) and those only described in the literature so far [1]. However, in a recent report Harmer [2] indicates that some of the commercial devices do not appear to be as accurate under intensive testing as would be hoped from the manufacturer's data. Hence there is still a need to consider techniques which can lead to the production of an inexpensive and relatively accurate fluorescence-based point temperature sensor. In this work, such a system will be described and its basis upon fluorescence from ruby induced by light from an LED can lead to its potential development as a high temperature device (>200°C) due to the crystalline nature of the transducer material.

PREVIOUS WORK

The authors have described previously fluorescence-based fibre optic temperature sensors based upon such materials as neodymium (in glass and YAG) and alexandrite [3]. However, these intensity-independent devices suffer from the small change in the parameter measured as a function of temperature (Nd), in spite of convenient infra-red excitation bands and the difficulty and expense of excitation with an electro-optically modulated laser (alexandrite). As a result, alternative materials which had equally good fluorescence characteristics, high temperature potential and ease of availability were sought. Ruby is such a material, fluorescing as it does in the region ~690nm due to the excitation of Cr^{3+} ions in the sapphire host. Sholes and Small discussed only **the principle** of such a device [4] with inconvenient xenon and tungsten lamp excitation and mechanical shutters, without the use of optical fibres. More recently, Bosselman et al [5] discussed a chromium doped lutetium - aluminium - chrome - borate material as a sensor element in an electronically self-oscillating system. These authors highlighted problems in the oscillator system and the wavelength separation of source and fluorescence.

PRINCIPLE OF OPERATION

The system relies upon the change in the decay characteristics of the transducer material with temperature. As the decay is exponential, to a close approximation, this method is an intensity - independent approach, thus obviating the need for intensity referencing. In previous work [3] this decay was monitored directly using short pulse excitation and accurate timing. It can be shown that the same information can be extracted as the 'phase-delay' between a sinusoidally modulated input and the emergent sinusoidally modulated fluorescence. The 'phase delay', ϕ is related to the 'decay-time', τ , by:

$$\tan \phi = \omega\tau$$

where ω is the frequency used. Such a technique offers considerable advantages when used with modern (simple) electronic devices - the information is carried by a wave of known modulation frequency and thus both the bandwidth of the electronic system can be limited and sensitive electronic filters may be used without loss of information. Ruby has a decay-time of ~3.5 ms and simple mathematics indicates that a suitable value of ϕ to yield a value of $\tan \phi > 1$ is several hundred Hertz. Digital signal averaging processes will be used and it can be shown a value of 125 Hz used is a good compromise between sensitivity and rapid evolution of data.

OPTICAL AND MECHANICAL DESIGN

A small piece of ruby of volume ~few mm^3 (cut from a ruby laser rod) is used. Light is coupled into and from the transducer element by two PCS1000 optical fibres. At this stage of the work, the probe is held together in a small metallic housing - in future a uv-cured adhesive bonded scheme will be employed. A single green LED (indicator type) is d.c. biased and sinusoidally modulated at 125 Hz. The peak emission at 565 nm (FWHM ~40 nm) couples well to the absorption peak in ruby. The exciting light is filtered optically by a narrow band interference filter centred at the peak wavelength (held firmly at 90° to the LED for constant optical characteristics). The Si p-i-n diode detector has a similar interference filter (peak wavelength ~692 nm, FWHM $\sim\pm$10nm).

Thus there is very high extinction of any residual 125 Hz modulated red light from the LED by the first filter and of the residual green light, (which mixes with the fluorescence (red) in the transducer), at the detector optics. Small changes in the LED spectrum e.g. due to current changes or ageing will thus only have a small effect on the intensity of the pumping (and not on the temperature dependent value of τ). The filters are chosen (with the 'cut-off' of the first being at a much shorter wavelength than the 'cut-on' of the second) so that small shifts in their transmission characteristics e.g. due to angle or temperature do not cause significant measurement errors in τ due to spillage of pump light from the LED. For accurate operation, it is important that these features are carefully considered at the design stage.

ELECTRONIC DESIGN

The principal function of the electronic aspects of the sensor system is the generation of the input frequency and detection of the received fluorescence in the presence of noise. Dedicated microprocessor techniques are used for signal averaging and the comparison of the measured phase lag, ϕ, with a calibration graph.

One important aspect of the design is the elimination of components which induce a phase-lag in the frequency detected, which is affected by factors other than the change in the fluorescence decay-time e.g. amplification factors and other component factors which are **not** independent of the signal intensity. Fixed phase changes in the electronic processing do not present a problem. A quartz crystal oscillator provides a stable 125 Hz signal both to drive the LED and provide a reference wave form. The detector is a Si p-i-n diode with an integral amplifier. This low bandwidth (~3kHz) component is both adequate and inexpensive. The signal if further amplified, then electronically filtered using an active filter. A commerical device ($Q = 10$) with optimal performance at 125 Hz was selected and the absence of a change in phase lag with input intensity level was confirmed. The signal is transformed to a square wave at the same frequency and the signal compared with the reference 125 Hz signal driving the LED. The phase-lag then is determined with respect to a 10MHz quartz oscillator and fed into a Z-80 microprocessor which is programmed to average 256 samples (in ~2s) and display the phase lag (and the standard deviation of the result sequentially). Alternatively the calibration graph may be stored and a direct display in degrees obtained.

RESULTS

Preliminary results of the calibration are reported. The probe is placed in an insulated oven, heated to ~170°C, and the cooling response determined. This is shown in Figure 1. A non-linear response is seen but a large phase-lag change with temperature results. The device is more accurate (at a modulation frequency of 125 Hz) at high temperatures where the slope of the calibration curve is greater - typically 3 standard deviations at 160°C corresponds to ~±0.25°C whilst this more than doubles at ~40°C.

Further results will be presented indicating the insensitivity to the intensity changes over a useful range and the potential for use in other temperature regions, especially high temperatures (T>200°C).

In summary, an accurate, inexpensive sensor has been produced using sophisticated yet simple electronics and signal processing combined with solid state optical and well known fluorescent materials technology. Negligible drift in signal with constant use is seen.

REFERENCES

1. Grattan K.T.V. Measurement and Control 20, 32-39, 1987.
2. Harmer A.L. Proc. 4th OFS (OFS 86) - Informal Workshop, Tsukuba Science City, Tsukuba, Japan, Pub: IECE, Japan; Tech. Digest ppVI1-VI8,1986.
3. Grattan K.T.V., Palmer, A.W., Rev. Sci. Instrum. 56, 1784-7, 1985. Augousti A.T., Grattan K.T.V., Palmer A.W. IEEE/OSA J. Lightwave Tech. LT5, 759-62, 1987.
4. Sholes R.R., Small, J.G. Rev. Sci. Instrum. 51, 882-4, 1980.
5. Bosselman T., Reule A., Schroder J. Proc. 2nd OFS (OFS84), Stuttgart, 1984.

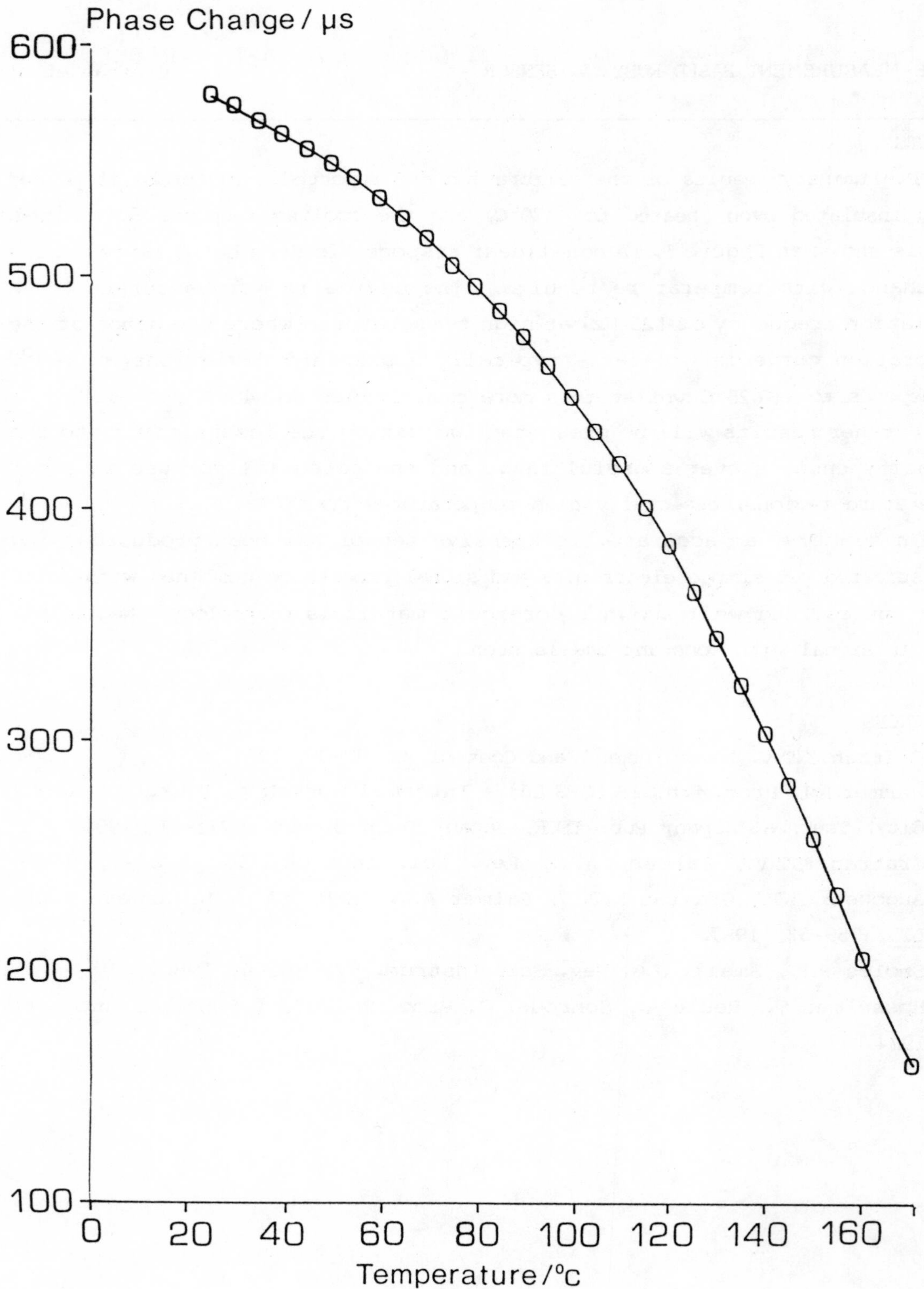

Figure 1: Calibration graph of the device.

SIMULTANEOUS MEASUREMENT OF TEMPERATURE AND PRESSURE VARIATIONS WITH A SINGLE MODE FIBER.

D. CHARDON, INSTITUT D'OPTIQUE , Bat 503, Centre Universitaire d'Orsay, BP 43, 91406 Orsay Cedex, FRANCE,

S. J. HUARD, ECOLE NATIONALE SUPERIEURE DE PHYSIQUE DE MARSEILLE, Domaine Universitaire de ST Jérome, 13397 Marseille Cedex 13, FRANCE.

Telephone : (33) 91 02 49 16 Telex : 402876F FACSTJE

INTRODUCTION

It is well known that the thermal sensitivity of an optical fiber sensor often appears as a parasitic effect. In the case of an intrinsic polarimetric pressure sensor, we have shown both theoretically and experimentaly [1] that the thermal sensitivity can be strongly reduced by performing a differential measurement. But in these experiments, the temperature variations are not available ; the pressure variations are deduced from the measurement of the induced birefringence (X and Y axes) in a single mode fiber wrapped around an hollow cylinder. When an overpressure is applied in the cylinder, the induced strains are transmitted to the fiber an so the refractive indices are affected by the photoelastic effect. In a similar manner, an increase of the temperature induces strains in the cylinder and modifies also the index of refraction of the fiber. Moreover, because of partial reflections on the fiber end-faces, such a device can be considered as a dual PEROT-FABRY resonator. Thus two optical parameters depending on P and T can be obtained : the mean round trip phase Φ and the differential phase shift Ψ and simultaneous measurement may be performed [2]. In this paper the theory of the operation is described and the first results on a pressure-temperature sensor are presented.

THEORY OF THE SIMULTANEOUS MEASUREMENT

Let us consider N turns of a single mode fiber coiled around an hollow cylinder with external radius R which can undergo internal pressure variations and homogeneous temperature variations. It has been shown [1] that the variations of the differential phase shift are directly proportionnal to the relative variations of curvature $\Delta R/R$. On the other hand, the variations of the mean round trip phase comprises two contributions ; the first one is proportionnal to $\Delta R/R$ and the second one, related to the thermooptic effect, is proportionnal to $\partial n/\partial T$. Finally when the pressure and the temperature are simultaneuosly modified, the situation can be discribed by the matrix relation :

$$\begin{pmatrix} \Delta\Psi \\ \Delta\Phi \end{pmatrix} = \begin{pmatrix} m_{11} & m_{12} \\ m_{21} & m_{22} \end{pmatrix} \begin{pmatrix} \Delta T \\ \Delta P \end{pmatrix}$$

where :

$$m_{21} = 4\pi NRk_0 \frac{\partial n}{\partial T} + m_{22} \cdot \frac{m_{11}}{m_{12}}$$

In the last expression k_0 is the wave number . These expressions clearly show that because of the thermooptic effect, the m_{ij} matrix is not singular and thus allows us to deduce both temperature and pressure variations from the measurements of $\Delta\Psi$ and $\Delta\Phi$ which must be available simultaneously. We have looked for a configuration which provides directly the differential phase shift rather than a separate mesurement of Ψ_x and Ψ_y [3] (fig. 1). More precisely, if the intensity transmitted through an linear analyzer, having an azimuth θ with respect the X-axis of the fiber coil, is evaluated it is possible to determine $\Delta\Psi$ and $\Delta\Phi$ .When the device is illuminated by a circularly polarised light the measurement of the transmitted intensities I_1, I_2, I_3, I_4, for θ equal 45°, 135°, 0°, and 90° gives the response.

The mean phase Φ and the differential phase Ψ can be deduced from the following relations :

$$\sin\Psi = \frac{\sqrt{\langle I_3\rangle.\langle I_4\rangle}}{2} \cdot \frac{I_1 - I_2}{I_3.I_4}$$

$$I_3 = \langle I_3\rangle . (1 + 2\rho \cos\Phi)$$

where <> is the mean value on one interference fringe and ρ the end-face reflection coefficient ($\simeq$ 0.04%).

EXPERIMENTAL SET-UP AND RESULTS

Five turns of a very low birefringence single mode optical fiber have been coiled on a 15 mm diameter hollow copper cylinder (1 mm thick). By bending the fiber becomes birefringent with its local principal axes parallel to the cylinder axis and normal to its surface. In order to observe simultaneously the interference phenomena due to the FABRY-PEROT configuration, a frequency stabilized He-Ne laser light is injected in the device. A stabilility better than 2 MHz has been obtained with an heated two longitudinal modes cavity [4]. An acousto-optic has been inserted between the laser and the fiber to prevent parasitic reflections in the cavity. The differents intensities needed for signal processing are delivered by four photodiodes illuminated through a polarizing set-up (fig. 2). The transmitted beam is separated in 4 channels by a polarizing prism in connection with an half-wave plate ; the plate-axes make a 22.5° angle with the prism axis. It is clear than the intensities I_1, I_2, I_3, I_4 are obtained by this system. The temperature is electrically controlled and the pressure can be adjusted inside the cylinder. A thermocouple and a piezo-electric manometer are the classical sensors. All the datas can be processed by a microcomputer which computes ΔT and ΔP from the intensities values.

The coefficients of the matrix have been first determined by applying a known variation on P or T. Without any ambiguity the matrix appears as a regular one ($m_{11}m_{22}-m_{12}m_{21}$ = 3.97 ± 0.25 rad^2 $°C^{-1}$ bar^{-1}). In order to measure simultaneously the temperature and the pressure variations, these quantities are modified by some degrees and by some bars. The optical measurement performed with our polarimetric-interferometric device agrees with the conventionnal measurement with an accuracy around 1%. The resolutions are respectively 0.1°C and 0.2 bar. The recording of T and P during the variations shows that the interference phenomena is not completely processed. Some crosstalk between the optical channels can explain this behaviour. Experiments are now in progress for a better isolation of these ones.

CONCLUSION

A new interferometric and polarimetric optical fiber sensor using a single mode fiber has been studied both theoritically and experimentally. It has been desmonstrated that simultaneous measurements of parameters could be achieved with a good accuracy. Pressure and temperature have been chosen for the demonstration but the principle can be applied with others parameters such as magnetic field, vibrations etc. This device appears as an interesting one when the temperature is a parasitic parameter in a measurement. The next study on this device will be an integration of the analysing system in order to keep a waveguide technology.

REFERENCES

[1] D. CHARDON, G. R. ROGER and S.J.HUARD ; OFS'86, TOKYO, October 1986.

[2] D.CHARDON, S.J.HUARD, R. KIST and W.OTT ; Europeen Patent 87104279.2.

[3] P.A. LEILABADY, J.D.C. JONES and D.A. JACKSON Opt. Lett. 10, (1985),576.

[4] P.E. CIDDOR and R.M. DUFFY ; J. Phys. E. Scient. Inst. 16, (1983), 1224.

FIGURES CAPTIONS

Fig. 1. Principle of the analyse of the light emerging from a birefringent FABRY-PEROT coiled fiber.

Fig. 2. Experimental device analysing the transmitted light by a the birefringent FABRY-PEROT coiled fiber.

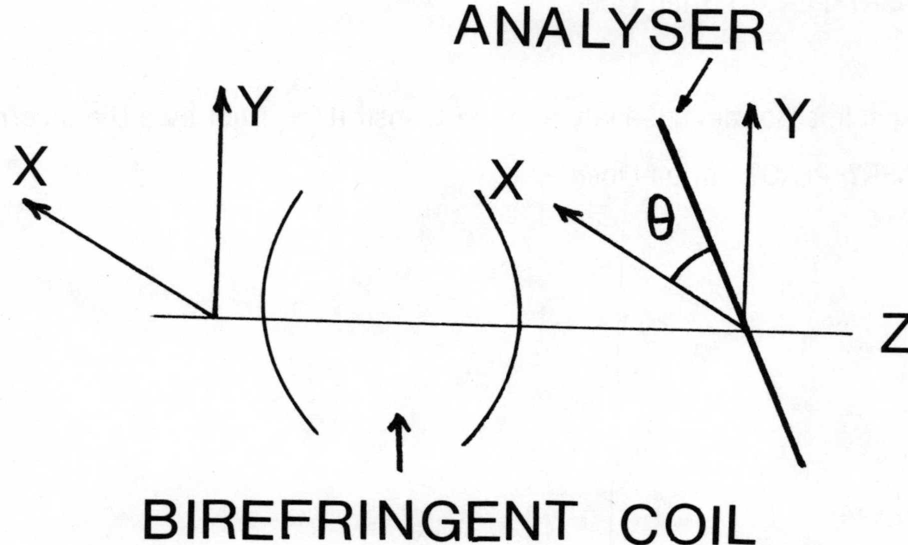
ANALYSER
Y
X
X
Y
θ
Z
BIREFRINGENT COIL

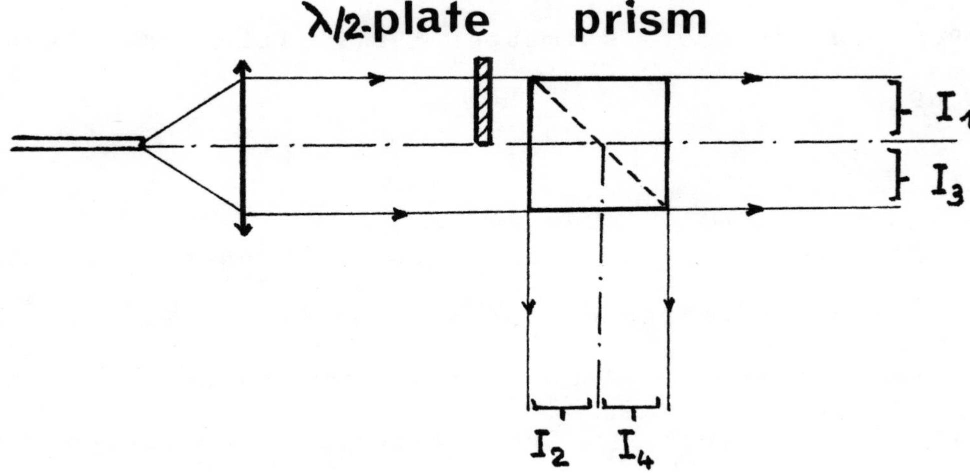
λ/2-plate
prism
I_1
I_3
I_2
I_4

NON-CONTACT TEMPERATURE MEASUREMENT WITH A ZIRCONIUM FLUORIDE GLASS FIBER

Serge MORDON, Elizabeth ZOUDE, Jean Marc BRUNETAUD

INSERM U279, 1 rue du Prof. Calmette, 59019 LILLE Cedex, FRANCE

(33) 20 52 69 20

SUMMARY

Zirconium Fluoride (ZF) glass fibers present a good transmission in the infrared region (1µm-5µm). We recently evaluated the ability to measure low temperatures without contact with ZF fibers. The principle considered for temperature measurement is based on the Planck distribution function describing the frequency spectrum of blackbody radiation. Our goal was to measure temperature from 60°C (337K) to 150°C (423K).

Two different ZF fibers (ZF1 and ZF2), from le Verre Fluoré (France) were used. ZF1 has a core diameter of 450µm and a cladding diameter of 550µm. ZF2 has a core diameter of 200µm and a cladding diameter of 250µm. These fibers have a step index structure. The losses between 1 and 4.5µm are below 1 dB/m and the numerical aperture is 0.18.

The infrared sensor is a lead selenide (PbSe) photoconductive detector from Optoelectronics, INC (USA). This detector is thermoelectrically cooled at -20°C (253K). The sensitivity is maximum at 4µm (2-5µm window). Then the detectivity is 2.E+10cmHz$^{1/2}$/W. The output signal of the detector is amplified by a lock-in amplifier (LIA). An optical chopper generates a reference signal. The infrared radiation is delivered by a

Temperature measurement with a Z.F. fiber. S. MORDON

blackbody. A lens is used to focus the infrared emission on the detector. LIA output signal is processed by a micro-computer in function of controlled blackbody temperature. Mean value, standard deviation, regression equation and correlation coefficient between experimental curve and mathematic curve are determined.

Results of the measurement are shown in fig.1. These results are obtained with fibers length of 1 m, fiber to blackbody distance equals 3cm and chopping frequency equals 1800Hz. LIA time constant is 1ms. For ZF1, regression equation is $S(v)=4.25E-9.T(°C)^{2.74}$ and correlation coefficient is .9993. For ZF2, regression equation $S(v)= 0.814E-9.T(°C)^{2.75}$ and correlation coefficient is .9992. The transmitted infrared signal is proportionnal to the core diameter. The ratio between LIA signal and surface is equal for each fiber. With a time constant of the demodulator equal to 300ms, the accuracy is +/- 0.2°C; for a time constant of 1ms, the accuracy is +/-2°C.

Applications of remote temperature measurement are important and concern temperature measurement of objects far from the observer or located in an inconvenient place. Wider use of this device can be predicted, including not only industrial but also medical and biological applications.

Temperature measurement with a Z.F. fiber. S. MORDON

Figure 1: LIA output voltage vs blackbody temperature.
Chopping frequency is 1800Hz. Time constant is 1ms.
ZF1: core diameter=450µm, fiber length=1m.
ZF2: core diameter=200µm, fiber length=1m.

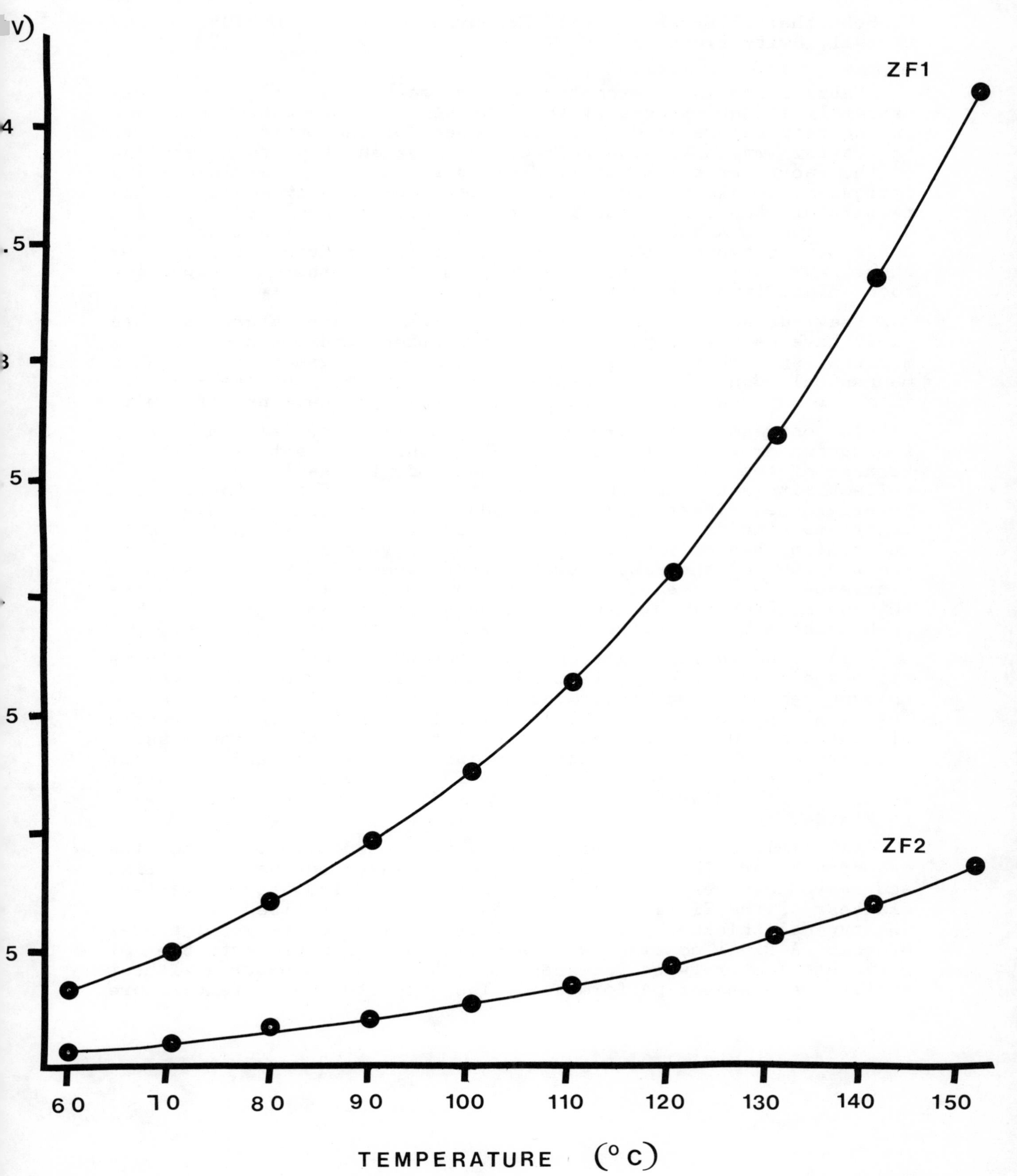
v)
4
.5
3
.5
5
5
ZF1
ZF2
60
70
80
90
100
110
120
130
140
150
TEMPERATURE (°C)

A SIMPLE FIBER-OPTIC FABRY-PEROT TEMPERATURE SENSOR

L. Schultheis, Brown Boveri Research Center, CH-5405 Baden/ Dättwil, Switzerland

Fabry-Perot interferometers are a sensitive tool to measure extremely slight changes of the optical cavity length. Variations in the temperature lead to small changes of the optical path i.e. the cavity length and the refractive index and thus to a detuning of the Fabry-Perot resonator. Whereas for most of the insulating optical materials the refractive index depends only weakly on the temperature the small bandgap semiconductors such as Si, GaAs, GaP, Ge etc. exhibit not only a high refractive index of about 3.5 but also typical thermal changes of the refractive index of about 10^{-4}/K which is much larger than the thermal expansion coefficient (typical in the 10^{-6}/K range)

Previous work on fiber-optic Fabry-Perot temperature sensors /1-4/ have mainly employed discrete optical components such as mirrors, spacers and lenses. Thus, careful alignment and stable mechanical mounts are necessary to optimize the Fabry-Perot interferometer which in turn renders many applications difficult.

In contrast, our approach is characterized by using a compact semiconductor platelet whose parallel surfaces act as mirrors. Because of the high refractive index of the semiconductor the reflectivity and thus the modulation depth of the Fabry-Perot interference fringes is high. In addition, the highly divergent light emerging from a typical multimode fiber and impinging onto the etalon has only a small internal divergence angle. Thus, smearing out of the Fabry-Perot interference due to the angular averaging is prevented for etalons of just a few microns thickness. Problems with collimation optics can be overcome. Two methods have been employed for fabricating the etalon sensor:

(i) Standard lapping techniques have been used to fabricate etalons of GaAs from a standard 0.5mm thick GaAs wafer. Typical thicknesses of about ten microns are achieved. After cleaving into small pieces of about 0.15mm x 0.15mm the etalon is mounted (1) between two fiber ends (in reflection <u>and</u> transmission geometry as schematically shown in inset of Fig. 1) and (2) just on one fiber tip in a simple reflection configuration. Mechanical stability is achieved by glueing the fiber and the etalon with polyimide.

(ii) Thin Si etalons have been evaporated directly onto the end of a fiber. The thickness of the Si layer can be controlled and reproduced very precisely. REM pictures indicate a perfect flatness of the Si film within 95% of the center fiber tip area. The typical thickness of a Si etalon is in the range of a few microns. A final coating with a polyimide film protects the Si platelet. Temperatures in excess of 300°C can be measured without spoiling the sensor performance. The absolute upper temperature limit is only given by the adhesion of the thin etalon on the fiber tip as well as the thermo-elastic properties of Si and fused silica and is expected to lie in the 600°C - 900°C range .

Fig. 1 depicts the transmitted and reflected intensity versus the photon energy of the GaAs etalon sensor. Conventional multimode fibers as well as fiber-optic connectors, couplers and a tungsten iodine lamp as a light source have been employed. The light is spectrally resolved by a grating monochromator (bandwidth 1nm) and detected by a photodiode. In the energy region below the bandgap strong Fabry-Perot interferences are seen, despite the fact that the reflected as well as the transmitted signal is averaged over the angular light distribution of the fiber. This observation confirms that the high refractive index of the etalon diminishes the effects of angular dispersion.

In Fig.2 we have plotted the reflected intensity versus the temperature of a Si etalon sensor (thickness is about 2500nm) for three different wavelengths. The spectral bandwidth is 1nm and the numerical aperture of the fiber is 0.2. The reflected intensity is uncorrected for a 4% Fresnel reflection from the second arm of the 3dB coupler which roughly accounts for the offset at the reflection minimum. A large modulation depth of the reflected intensity is observed because for the thin Si etalon effects arising from the angular averaging are completely negligible (in contrast to the thick GaAs etalon). The maximum temperature resolution of this Si etalon sensor is about 3°C and is limited by intensity fluctuations (modal noise) in the fiber.

The resolution, in principle, can be easily improved by chosing a thicker etalon and thus a higher etalon sensitivity. However, the reflection modulation depth decreases because of the angular averaging for the case of a high aperture multimode fiber. In addition, the temperature range in which the Fabry-Perot reflectance gives the temperature without any ambiguity also decreases necessitating an electronic phase tracking system to determine the temperature over the full temperature range.

A second approach to increase the resolution of the sensor is to use the reflectivity from the etalon at two different wavelengths. This can be done either by using two light sources or by spectrally modulating one light source. These schemes for internal referencing would solve the problems arising from modal noise as well as perturbations arising from the handling of the fiber.

In conclusion, we have investigated a fiber-optic semiconductor etalon temperature sensor. The etalon sensor exhibits a large versatility of the sensor design (semiconductor, thickness, numerical aperture of the fiber) concerning the temperature range and the sensitivity. In particular, simple batch fabrication techniques for the etalon sensor such as evaporation or sputtering offer the potential for low cost economic temperature sensing.

REFERENCES

1. D.A. Christensen, 1974 Annual Meeting of the Optical Society of America
2. E.R. Cox and B.E. Jones, Optical Fibre Sensors, IEE 221, 122 (1983)
3 K.A. James, W.H. Quick and J.E. Coker, Optical Fibre Sensors, IEE 221, 6 (1983)
4 G. Boreman, R. Walter, and D. Lester, Fiber Optics and Laser Sensors III, SPIE vol. 566, 312 (1985)

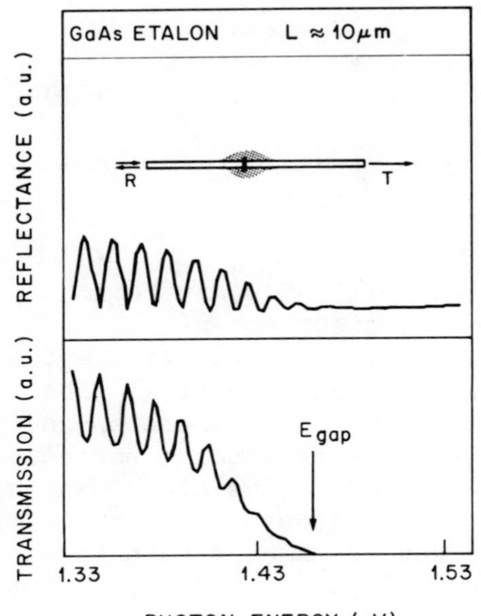

Fig.1: Reflected and transmitted intensity versus photon energy in the energy region below the bandgap E_g of a GaAs etalon sensor at room temperature

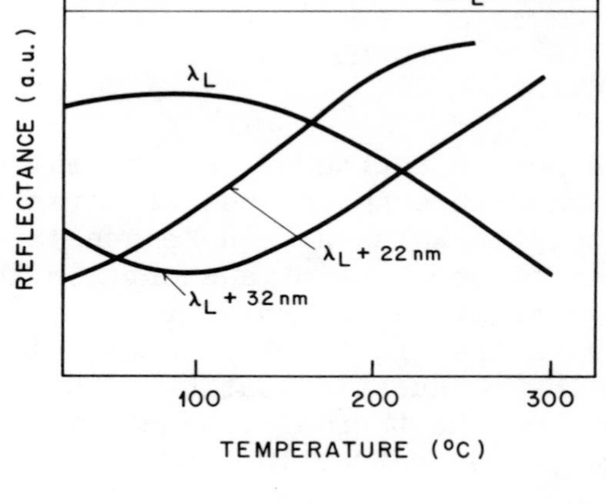

Fig.2: Reflected intensity versus temperature of a Si etalon sensor for three different wavelengths

FFF5-1

AN OPTICAL FIBER THERMAL CONDUCTIVITY SENSOR

B. J. White, J. P. Davis, L. C. Bobb
Naval Air Development Center
Warminster, PA 18974

D. C. Larson
Drexel University
Philadelphia, PA 19104

INTRODUCTION

The determination of the thermal conductivities of liquids may be accomplished through the use of a variation of the transient hot-wire technique,[1] in which a Mach-Zehnder interferometer is employed to measure the change in temperature of a short length of conductively coated optical fiber.[2] This segment of fiber, which forms part of one arm of the interferometer and is immersed in the liquid of interest, is heated resistively by passing current through the conductive coating. The increase in the fiber temperature, which depends upon the thermal conductivity of the surrounding liquid, results in both a change in the refractive index of the fiber core and an axial strain in the fiber and is observed as a shifting of the interference pattern at the output of the interferometer. This shifting of the interference fringes may be directly correlated to the temperature change, and the thermal conductivity determined from the rate of change of the temperature.

FIBER TEMPERATURE SENSITIVITY

A theoretical determination of the temperature sensitivity of the fiber may be made from a consideration of various fiber parameters and the assumption of uniform fiber temperature at each time.[2,3] The result for the ITT single-mode fiber used in this study is

$$\frac{\Delta\phi}{L\Delta T} = 102.5\ \frac{\text{rad}}{\text{m}\cdot\text{K}} = 16.3\ \frac{\text{fringes}}{\text{m}\cdot\text{K}} \qquad (1)$$

in which $\Delta\phi$ is the phase change, L is the length of fiber heated, and ΔT is the temperature change. Because of approximations in the theoretical model and uncertainties in the values of some of the parameters used in this determination, however, this value of the temperature sensitivity does not agree precisely with that obtained experimentally. Both our research and that performed in a study by Lagakos et al.,[4] in which they used a furnace to heat the fiber in one arm of a Mach-Zehnder interferometer, indicate a value of the temperature sensitivity 2.7 to 2.9% lower than the theoretical value. Hence, the temperature changes reported in this summary were calculated from the number of fringes produced, the length of fiber being heated, and the experimental temperature sensitivity

$$\left(\frac{\Delta\phi}{L\Delta T}\right)_{exp} = 15.9\ \frac{\text{fringes}}{\text{m}\cdot\text{K}} \qquad (2)$$

EXPERIMENTAL RESULTS

The experimental configuration, illustrated in Figure 1, included a Mach-Zehnder interferometer, which was fabricated from two ITT fused couplers. Light from a He-Ne laser was focused onto the core of the input fiber and was collected by a photodiode at the output. The test fibers, which had been sputtered with a thin layer of gold along a small portion of their length, were fitted with TRW Optalign connectors to permit easy insertion into the sensing arm of the interferometer. The interferometer output was recorded on a Gould ES 1000 recorder. In this manner, it was possible to measure the actual phase shift (by counting the number of fringes produced) per unit temperature change caused by the application of current to the conductive coating.

A series of experiments was performed in which square-wave voltages of different magnitude were applied across gold-coated optical fibers immersed in glycerol, ethylene glycol, and water. The change in the fiber temperature as a function of time was determined by monitoring the number and spacing of the interference fringes produced. The temperature change can be determined at any time by dividing the number of fringes produced until that time by the value in Eq. (2) and by the length of fiber being heated.

The experimental data were compared with the solution of the conduction equation given by Carslaw and Jaeger[5] for a cylinder of radius a and infinite thermal conductivity surrounded by an infinite medium and heated at the rate Q per unit length per unit time, t. The solution for large τ, where $\tau = kt/\rho c_p a^2$, k is the thermal conductivity of the surrounding medium, ρ is the density, and c_p is the heat capacity, is approximately

$$\Delta T = \frac{Q}{4\pi k}\left\{\left[\ln \frac{4\tau}{C}\right]\left[1 + \frac{\alpha_1 - 2}{2\alpha_1 \tau}\right] + \frac{1}{2\tau}\right\} \qquad (3)$$

in which C=1.7811 and $\alpha_1 = 2(\rho c_p)_{fluid}/(\rho c_p)_{fiber}$. For sufficiently large values of the time, the last two terms in Eq. (3) become negligible so that a curve of ΔT vs. $\ln \tau$ has a linear asymptote of slope Q/k, and thus, if Q is known, k may be determined.

For each liquid, the temperature change was determined as a function of time t for each of several power levels applied to the fiber's conductive coating. These time-dependent changes in temperature were then normalized by dividing each by its respective power per unit length, Q. Since τ is dependent upon k, we plotted $\Delta T/Q$ versus ln t, where t is measured in seconds. These curves were consistently reproducible for the different power levels attempted. This is evidenced in Figure 2, which shows data sets obtained at three different powers while a 4.0-cm-long segment of optical fiber was immersed in ethylene glycol.

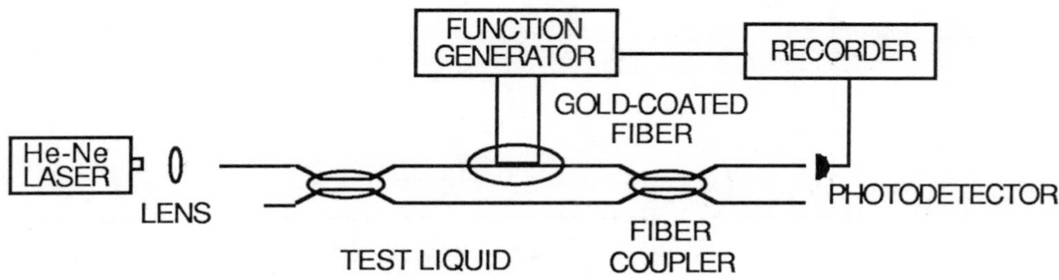

Figure 1. Optical-fiber Mach-Zehnder interferometer.

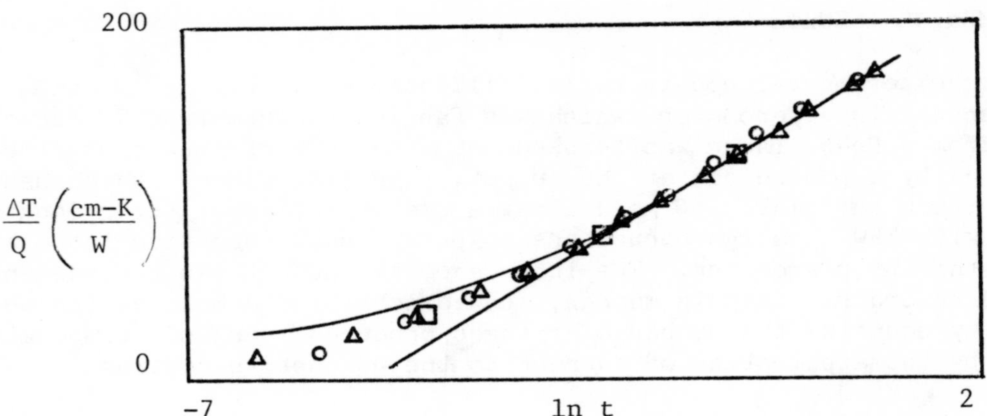

Figure 2. Plot of the normalized temperature change as a function of the natural logarithm of the time for a 4.0-cm-long fiber in ethylene glycol. The symbols represent the experimental data points obtained at 0.159 W (squares), 0.436 W (circles), and 0.512 W (triangles). A value of $k=2.80\times10^{-3}$ W/cm·K was used in fitting the theoretical curve and its linear asymptote (solid curves) to the data.

A curve derived from the theoretical conduction model is shown along with its linear asymptote for large t and the ethylene glycol data of Figure 2. The final slope of the theoretical curve may be made approximately equal to that of the experimental data by adjusting the value of k in Eq. (3). A value of 2.80×10^{-3} W/cm·K was obtained in this manner for the thermal conductivity of ethylene glycol. This value is somewhat higher than the accepted thermal conductivity[6] at 300 K of 2.576×10^{-3} W/cm·K, but still within the range of values obtained by other investigators. (See Table 1.) The values of k we obtained at different power levels were consistently high and may be indicative of the presence of an impurity, such as water, in the sample.

The same procedure was used to measure the thermal conductivities of glycerol and water. Whereas the heat lost from the optical fiber when situated in ethylene glycol and glycerol was primarily by conduction, the determination of a value of k for water was complicated by the presence of a second mechanism of heat loss, which we believe to be convection. As shown in Figure 3, the experimentally determined values of k, obtained by fitting Eq. (3) to the data,

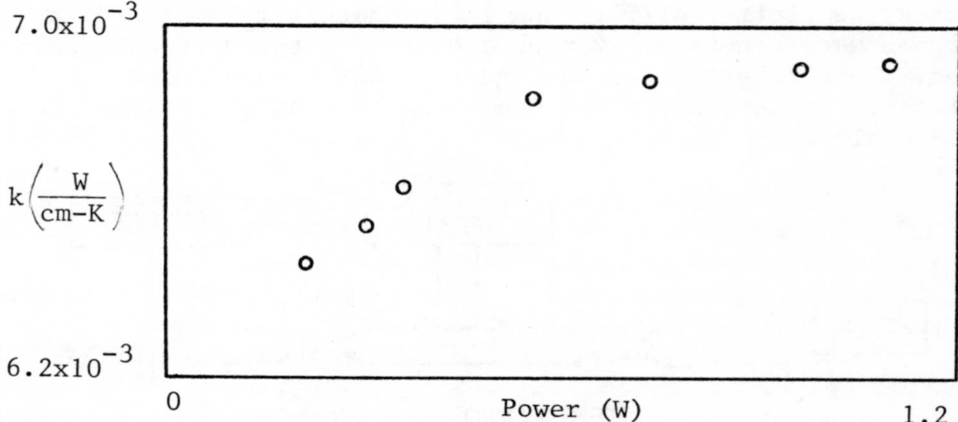

Figure 3. Experimentally determined values of the thermal conductivity of water as a function of power to the fiber.

Table 1. Summary of Results of Thermal Conductivity Measurements

Liquid	Experimental k (mW/cm·K)	Accepted[6] k (mW/cm·K)	Range of accepted values	Deviation of present experiment from accepted value of k
Ethylene glycol	2.80	2.576	-7% to +10%	+8.7%
Glycerol	2.93	2.880	-2% to +3%	+1.7%
Water	6.21	6.084	-2% to +3%	+2.1%

increased with the power to the fiber coating. The first four points on the graph fall roughly on a line, which, when extrapolated to zero power (to minimize convective loss), yields a value of 6.21×10^{-3} W/cm·K for the thermal conductivity.

A table summarizing our results and comparing them with the recommended values[6] of the thermal conductivity is presented above. The dispersion in the values at 300 K found by the various investigators cited in Ref. 6 is indicated for each liquid. As shown, all of the values of k obtained in the present study are within the acceptable ranges.

CONCLUSIONS

We have demonstrated that thermal conductivities of liquids may be determined through the use of an optical-fiber interferometric technique in which the optical fiber acts as both the heating element and the thermometer. Good agreement was obtained between our experimental results and published values of the thermal conductivities of ethylene glycol and glycerol. The experiments performed with the fiber in water yielded progressively higher values of k as the power to the fiber was increased and indicate that care should be taken when employing this technique to avoid or to compensate for the introduction of convective motion in the liquid.

REFERENCES

1. N. Fox, N.W. Gaggini, and R. Wangsani, "Measurement of the thermal conductivity of liquids using a transient hot wire technique," Am. J. Phys. **55**, 272 (1987).
2. B. J. White, J. P. Davis, L. C. Bobb, H. D. Krumboltz, and D. C. Larson, "Optical-fiber thermal modulator," J. Lightwave Tech. **LT-5** (Sept. 1987).
3. L. S. Schuetz, J. H. Cole, J. Jarzynski, N. Lagakos, and J. A. Bucaro, "Dynamic thermal response of single-mode optical fiber for interferometric sensors," Appl. Opt. **22**, 478 (1983).
4. N. Lagakos, J. A. Bucaro, and J. Jarzynski, "Temperature-induced optical phase shifts in fibers," Appl. Opt. **20**, 2305 (1981).
5. H. S. Carslaw and J. C. Jaeger, **Conduction of Heat in Solids**, Oxford: Oxford University Press, 1959.
6. Y. S. Touloukian, "Thermal conductivity -- nonmetallic liquids and gases," Vol. 3 of **Thermophysical Properties of Matter**, The TPRC Data Series, New York: IFI/Plenum, 1970.

KEY TO AUTHORS AND PAPERS